The Scholarship of Teaching and Learning

THE SCHOLARSHIP OF TEACHING AND LEARNING

A Guide for Scientists, Engineers, and Mathematicians

JACQUELINE M. DEWAR
Loyola Marymount University, Los Angeles, CA, US

CURTIS D. BENNETT
California State University, Long Beach, CA, US

MATTHEW A. FISHER
Saint Vincent College, Latrobe, PA, US

OXFORD
UNIVERSITY PRESS

Great Clarendon Street, Oxford, OX2 6DP,
United Kingdom

Oxford University Press is a department of the University of Oxford.
It furthers the University's objective of excellence in research, scholarship,
and education by publishing worldwide. Oxford is a registered trade mark of
Oxford University Press in the UK and in certain other countries

© Oxford University Press 2018

The moral rights of the authors have been asserted

First Edition published in 2018

All rights reserved. No part of this publication may be reproduced, stored in
a retrieval system, or transmitted, in any form or by any means, without the
prior permission in writing of Oxford University Press, or as expressly permitted
by law, by licence or under terms agreed with the appropriate reprographics
rights organization. Enquiries concerning reproduction outside the scope of the
above should be sent to the Rights Department, Oxford University Press, at the
address above

You must not circulate this work in any other form
and you must impose this same condition on any acquirer

Published in the United States of America by Oxford University Press
198 Madison Avenue, New York, NY 10016, United States of America

British Library Cataloguing in Publication Data
Data available

Library of Congress Control Number: 2017957047

ISBN 978-0-19-882121-2

Links to third party websites are provided by Oxford in good faith and
for information only. Oxford disclaims any responsibility for the materials
contained in any third party website referenced in this work.

FOREWORD

"It is our awareness of being unfinished that makes us educable."

Paulo Freire, *Pedagogy of freedom: Ethics, democracy, and civil courage*

Over a decade ago, I found myself having an informal but deeply engaging chat with a good colleague at the annual conference of the International Society for the Scholarship of Teaching and Learning (ISSOTL). My colleague, who is a professor of history and law, and I were intrigued by how disciplinary assumptions and approaches to inquiry are embedded from the very early stages of our undergraduate students' epistemological development. I have been reflecting back on this conversation (which actually continued over the course of several years), as I immersed myself in the chapters of Jacqueline Dewar, Curtis Bennett, and Matthew Fisher's present volume.

The scholarship of teaching and learning has allowed us to make visible the richness and complexity of our disciplinary and inter-, multi-, or cross-disciplinary understandings of teaching and learning. As a scientist, my SoTL journeys have taken me into delightful "unfamiliar territory," (Reichard and Takayama 2012) as I ventured into collaborations with humanists, social scientists, and artists. Such intersections have given me the opportunity to interrogate my assumptions about student learning. Yet, to be able to fully appreciate the insights gained from these intersections, those new to the scholarship of teaching and learning benefit first from unpacking their own disciplinary expertise (see Pace and Middendorf 2004) and signature pedagogies (Shulman 2005) in order to shape a path of inquiry that originates on familiar ground. But how to start?

The Scholarship of Teaching and Learning: A Guide for Scientists, Engineers, and Mathematicians is the book I wish I had as a faculty member venturing into SoTL many years ago. Dewar, Bennett, and Fisher have drawn on their own developmental process in SoTL as STEM faculty, and their experiences facilitating workshops and learning communities for the STEM community, to bring their collective expertise into a clear and compelling volume. The authors define SoTL as "the intellectual work that faculty members do when they use their disciplinary knowledge to investigate a question about their students' learning (and their teaching), gather evidence in a systematic way, submit their findings to peer review, and make them public for others to build upon" (Chapter 1, The Origins of SoTL in Higher Education). Their approach is inclusive and accessible, and reassuring in its stance of SoTL not as an end in itself or the primary motivation for inquiry, but as our openness to critique—indeed, our own cognizance of our "unfinishedness"—to allow others to build upon this work.

The authors are respectful of scientists' natural tendency to define variables and design controlled experiments. They translate to the STEM community the process of problematizing teaching in relation to disciplinary knowledge, situational factors, and most importantly, our students' learning. The methodologies of SoTL are clearly explained and STEM scholars will appreciate the explanations of the standards used to assess quantitative and qualitative research. I was impressed with the diverse and representative examples of instruments, taxonomies, and models for SoTL studies in STEM that were collated in Appendix IV. STEM faculty will appreciate the immediate applicability of these resources to their disciplines. Each chapter progressively scaffolds the process for inquiry, which is engagingly rich for those new to SoTL, but also invaluable for professionals in teaching and learning centers who develop and facilitate SoTL workshops and institutes.

Importantly, the authors carefully and considerately discuss the role of SoTL as research in tenure and promotion decisions by drawing on their own collective trajectories as faculty, as institutional leaders, and their experiences with disciplinary societies, together with their survey of cases across institutional and international contexts.

I am grateful to my colleagues for writing this volume, which greatly strengthens our commitment to teaching as community property (Shulman 1993) in the STEM disciplines and beyond. As I revisit my disciplinary assumptions to yet again be confronted and delighted with my ongoing "unfinishedness," I hope that you will enjoy this book as much as I have.

Kathy Takayama, PhD
Former President, International Society for the Scholarship
of Teaching and Learning
Director, Center for Advancing Teaching and Learning
Through Research, Northeastern University

REFERENCES

Pace, D., and J. Middendorf, editors. 2004. *Decoding the Disciplines: Helping Students Learn Disciplinary Ways of Thinking*, New Directions for Teaching and Learning: No. 98. San Francisco: Jossey-Bass.

Reichard, D., and K. Takayama. 2012. "Exploring Student Learning in Unfamiliar Territory: A Humanist and a Scientist Compare Notes." In *The Scholarship of Teaching and Learning In and Across the Disciplines*, edited by K. McKinney, pp. 169–85. Bloomington: Indiana University Press.

Shulman, L. S. 1993. "Teaching as Community Property: Putting an End to Pedagogical Solitude." *Change 25* (6): pp. 6–7.

Shulman, L. S. 2005. Signature Pedagogies in the Professions. *Daedalus 134* (3): pp. 52–9.

PREFACE

As a science, technology, engineering, or mathematics (STEM) faculty member in higher education who cares deeply for your students, you have probably had long talks with colleagues about students, read books on teaching and learning, and attended events at your campus's teaching and learning center. You may have read journal articles or attended conference talks that reported the results of scholarly studies of teaching and learning. Now you may be thinking about taking the next step, investigating a question you have about your teaching and your students' learning, that is, engaging in a scholarship of teaching and learning (SoTL) project of your own. Such an undertaking may seem daunting, particularly if you have no formal training in studying teaching and learning. These observations described each of the three authors of this book at some point in the past. Each of us earned advanced degrees in our respective STEM disciplines and went on to become faculty members at American universities. Each of us was deeply interested in our students and their learning, and we had no formal training in studying teaching and learning. Today, all three of us can say that studying teaching and learning has enriched our professional lives.

We were formally "inducted" into the SoTL community, when we were selected as Carnegie Academy for the Scholarship of Teaching and Learning (CASTL) scholars—Bennett in 2000-01 and again in 2003-04, Dewar in 2003-04, and Fisher in 2005-06. Carnegie's CASTL Scholar Program afforded us opportunities to work with scholars from diverse academic fields, including biology, chemistry, computer science, physics, mathematics, engineering, and health sciences. This experience prepared and inspired us to promote SoTL by presenting "how-to-do-SoTL" workshops and mentoring faculty in STEM and other disciplines on campuses and at conferences across the US and in Canada. In doing these workshops we learned that, when STEM faculty begin to do

SoTL, they generally bring similar knowledge, strengths, and needs that diverge somewhat from those of non-STEM faculty. That observation led to this how-to guide for STEM faculty interested in pedagogical research as well as for those engaged in work related to STEM education grants, assessment, or accreditation. We have written with the needs of those teaching in the computer, earth, environmental, health, life, and physical sciences, engineering, and mathematics in mind, but we believe that economists and social scientists interested in SoTL would also find this book helpful.

This book takes readers through the process of designing, carrying out, and publishing a study from beginning to end, using many examples of studies of teaching and learning in STEM. To guide them through the process, we periodically interrupt the narrative to insert questions that prompt readers to reflect, plan, or act. These locations are clearly marked as places to "Pause."

We have aimed this book at an international audience. However, sometimes space or other considerations limited our discussions to the United States (US), for instance, when we described the origins of STEM education research in Chapter 1. Language variations also presented us with some challenges; for example, the most appropriate way to refer to college or university level instructors differs by country. It might be instructor, teacher, professor, faculty, or faculty member. We have used these interchangeably and trust that readers will understand and accept our decision on this matter when they encounter one that seems odd to them.

The book includes a detailed Table of Contents as well as an Index.

Chapter 0 explains why the scholarship of teaching and learning should be of interest to faculty in science, engineering, or mathematics at all types of institutions, including faculty members active in traditional research.

Chapter 1 describes the origins of the scholarship of teaching and learning movement, explores the distinctions among SoTL, good teaching, scholarly teaching, and discipline-based education research (DBER) on undergraduate teaching and learning in STEM fields. We provide an overview of the origins of educational research in the US in physics, biology, chemistry, and engineering, and then examine in more depth mathematics as a case study. The chapter addresses the issue of evaluating and valuing this work, including implications for junior faculty who wish to engage in SoTL.

Chapter 2 shows how to convert a teaching "problem" into a researchable question. It presents a taxonomy of SoTL questions derived from the work of Carnegie scholars that we have found useful in guiding the development of a project and shows how disciplinary knowledge, situational factors, and exploring assumptions can be brought to bear on framing SoTL research questions. It also describes how to conduct a search of educational literature.

Chapter 3 examines basic considerations of education research design, such as whether or not to have experimental and control groups. It discusses the need to obtain human subjects clearance in order to publish the results of a SoTL study, a possibly unfamiliar topic for many STEM faculty members.

Chapter 4 provides an introduction to gathering data for SoTL investigations, including the importance of triangulating data, examples of different types of quantitative data and qualitative data, and how to design assignments so that the coursework students produce can also serve as evidence for a SoTL study.

Chapter 5 describes the use of surveys in SoTL studies, including a special type of survey called a knowledge survey.

Chapter 6 gives detailed instructions for methods of gathering evidence that may be unfamiliar to STEM instructors, such as focus groups, interviews, and think-alouds.

Chapter 7 focuses on methods for analyzing qualitative data such as rubrics and content analysis or coding. It also compares and contrasts the standards used to assess quantitative and qualitative research.

Chapter 8 provides information on finding collaborators, sources of support, possible venues for dissemination, getting a manuscript written and submitted, and responding to reviews or rejection.

Chapter 9 presents the authors' views on the value of SoTL to faculty, to their institutions, to the academy, and to students.

The four appendices contain a list of the Carnegie CASTL scholars in STEM and related fields; a list of the participants in the Carnegie Scholarly and Professional Societies Program; an example of a focus group protocol; and a list of surveys, scales, and taxonomies that might be useful for SoTL studies in STEM fields.

ACKNOWLEDGMENTS

We are grateful to our editor Dan Taber for his unfailing support and encouragement throughout the entire process of bringing this book from just an idea to actual publication and to the OUP reviewers of our manuscript for suggesting improvements to this book. We want to acknowledge the significant role that Carnegie President Lee Shulman and Vice President Pat Hutchings, our Carnegie mentors, Marcia Babb, Richard Gale, Mary Huber, and all of our Carnegie scholar colleagues—but most especially, Spencer Benson, Kate Berheide, Pat Donahue, Jose Feito, Kathleen McKinney, David Reichard, Anita Salem, and Whitney Schlegel—played in helping us develop as scholars. We are indebted to Carnegie scholars Brian Coppola and Michael Loui, who suggested many good examples of published SoTL studies in science and engineering. A special thank you goes to Carnegie scholar Kathy Takayama for writing the Foreword and to the following individuals who responded to our requests for specific information about the scholarship of teaching and learning in their countries: Carnegie scholars Vaneeta D'Andrea (England), David Geelan (Australia), and Bettie Higgs (Ireland); and SoTL colleagues Nancy Chick (Canada), James Cronon (Ireland), and Joelle Fanhanel (England).

CONTENTS

0. A Worthy Endeavor	1
1. Understanding the Scholarship of Teaching and Learning (SoTL)	5
Introduction	5
The Origins of SoTL in Higher Education	6
Distinguishing SoTL from "Good" and "Scholarly" Teaching	8
Forging Disciplinary Connections for SoTL	9
Discipline-Based Education Research (DBER)	10
SoTL and DBER: A Case Study in Mathematics	11
Evaluating SoTL	15
Valuing SoTL: Special Considerations for Junior Faculty Members	17
2. Developing a Researchable Question	22
Introduction	22
A Taxonomy of SoTL Questions	23
Generating Questions from Problems	26
Narrowing Questions	27
Grounding a Question	28
Identifying Assumptions	33
Feasibility	34
Revisiting Goals	35
3. Designing a Research Study	39
Introduction	39
Challenges of Education Research Design for SoTL	39
Human Subjects Considerations	51
4. Gathering Evidence: The Basics	60
Introduction	60
Triangulating Data	60
Quantitative Versus Qualitative Data	61
Familiar Sources of Evidence	62
Assignment Design	65

5. **Evidence: From Surveys** — 68
   - Introduction — 68
   - Surveys — 68
   - Knowledge Surveys — 75

6. **Evidence: From Interviews, Focus Groups, and Think-Alouds** — 86
   - Introduction — 86
   - Interviews — 86
   - Focus Groups — 92
   - Think-Alouds — 97
   - Student Voices in the Scholarship of Teaching and Learning — 105

7. **Analyzing Evidence** — 111
   - Introduction — 111
   - Rubrics — 113
   - Coding Data (Content Analysis) — 120
   - Assessing the Quality of Content Analysis — 126

8. **The Final Step for Doing SoTL: Going Public** — 132
   - Introduction — 132
   - Finding Collaborators and Support — 132
   - Presenting a Paper — 134
   - Publishing an Article — 137

9. **Reflecting on the Benefits of the Scholarship of Teaching and Learning** — 145

Appendix I: Carnegie CASTL Scholars in STEM and Related Fields — 151

Appendix II: Participants in the Carnegie Scholarly and Professional Societies Program — 159

Appendix III: Sample Focus Group Protocol — 161

Appendix IV: Instruments, Taxonomies, and Models for SoTL Studies in STEM Fields — 170

Index — 175

CHAPTER 0

A Worthy Endeavor

The history of higher education offers many examples of scientists, engineers, and mathematicians who began their career as researchers and later became involved in improving education. In 1900, the great mathematician Felix Klein wrote thoughtfully about pedagogy and was elected the first president of the International Commission on Mathematics Instruction. Around the same time, in biology, Franklin Paine Mall, the first head of anatomy at Johns Hopkins University School of Medicine made major reforms in the teaching of anatomy (Buettner 2007). Somewhat before that, engineers like Robert H. Thurston at Cornell initiated a debate that took decades to resolve about the proper balance in engineering curricula between preparation for real-world practice and involvement in research (Marcus 2005).

While these scientists, engineers, and mathematicians of the late 1800s and early 1900s wrote about how and what to teach, they were not engaged in what we would consider the scholarship of teaching and learning (SoTL) or discipline-based education research. By the 1970s, however, active disciplinary researchers were making contributions to pedagogical research in science, engineering, and mathematics. In addition to his significant work in pure mathematics, Hans Freudenthal of the Netherlands established what is now the Freudenthal Institute for Science and Mathematics Education (O'Connor and Robertson 2000). Miles Pickering, earned a doctorate in chemistry from State University of New York, Stony Brook, and worked at Princeton prior to becoming a well-known scholar in chemistry education (Kandel 1997). Richard Felder of North Carolina State University published widely in chemical engineering before shifting the focus of his research to engineering education. More recently, Carl Wieman, recipient of the 2001 Nobel Prize in Physics, became a leading

The Scholarship of Teaching and Learning, Jacqueline M. Dewar, Curtis D. Bennett, and Matthew A. Fisher.
© Oxford University Press, 2018. Published 2018 by Oxford University Press

researcher in the teaching and learning of physics (Deslauriers et al. 2011; Hadfield and Wieman 2010; Holmes et al. 2015; Smith et al. 2009) and a fervent advocate for changing how science is taught (Wieman 2007, 2017).

An informal community of SoTL scholars in science, technology, engineering, and mathematics (STEM) arose in higher education in the late 1990s. This community tends to engage in a more action-based research, that is, research where professors investigate the learning in their own classrooms in a scholarly fashion. This work requires time, and in today's world of specialization, many will argue that most faculty members would be more productive focusing on their disciplinary research. We believe, however, that wide participation in SoTL, including faculty active in traditional research, benefits the entire teaching and learning enterprise. Undertaking both traditional research and SoTL is not easy. But we have found that faculty involved in traditional disciplinary research bring important and different perspectives to SoTL.

The same can be said for STEM faculty at community colleges, but for different reasons. While they face little or no expectation for scholarly research, their duties include significantly greater teaching assignments, often with courses at the most elementary levels and students who are older and have more life experience. Their situation offers a rich context for study and research (Burns 2017; Huber 2008; Tinberg et al. 2007).

While some faculty engaged in pedagogical research may spend significant time learning to perfect the techniques used in such research (many of which are discussed in this book), making contributions to teaching and learning in our disciplines does not always require complete perfection. We should not let the perfect be the enemy of the worthwhile. This version of an old aphorism nicely sums up a point-of-view shared with us by Carnegie scholar Kate Berheide. In fact, engaging in pedagogical research is one of the best ways to learn more about doing it.

No one else is likely to study our question in our classroom. Each small study makes a contribution to the collective knowledge about teaching and learning in science, engineering, and mathematics and supports the effort to make teaching in these disciplines community property (Shulman 1993).

Speaking from our experience, doing SoTL can be rewarding for anyone in the professoriate. Investigating teaching and learning provides us with many benefits, ranging from improving our own teaching and making it more enjoyable to offering the intellectual challenge of trying to develop better pedagogical approaches. The scholarship of teaching and learning helps us avoid what Shulman (1999) described as "pedagogical amnesia," the many things about our teaching we forget from one semester to the next. SoTL enables us to capture insights into our teaching in a way that allows us to revisit them time and again.

SoTL work redirects our attention from whether or not students are learning topics to understanding the reasons why students have difficulty learning those topics. Consequently, SoTL gives us greater insights into teaching and learning, which moves us toward the ultimate goal of improving student learning.

Perhaps most importantly, SoTL makes explicit the intellectual challenges involved in teaching. SoTL spurs us to discover what happened when a course goes poorly, whether because of disappointing student performance, negative student evaluations, or something else entirely. The process of investigation encourages us to confront our assumptions about students and their learning and to try to determine whether or not they are valid. Learning can be far more complex than we might imagine. Collecting and analyzing evidence to gain insight into the factors that inhibit or promote learning can be intellectually stimulating. Using our content knowledge to discover how to help students overcome these difficulties is a challenging but worthy endeavor. Making the effort rewards us with increased satisfaction in teaching, greater faith in students, and, in the end, a real sense of accomplishment.

REFERENCES

Buettner, K. 2007. "Franklin Paine Mall (1862–1917)." *Embryo Project Encyclopedia*. Accessed August 31, 2017. http://embryo.asu.edu/handle/10776/1682.

Burns, K. 2017. "Community College Faculty as Pedagogical Innovators: How the Scholarship of Teaching and Learning (SoTL) Stimulates Innovation in the Classroom." *Community College Journal of Research and Practice 41* (3): pp. 153–67. doi: 10.1080/10668926.2016.1168327.

Deslauriers, L., E. Schelew, and C. Wieman. 2011. "Improved Learning in a Large-Enrollment Physics Class." *Science 332* (6031): pp. 862–4. doi: 10.1126/science.1201783.

Hadfield, L., and C. Wieman. 2010. "Student Interpretations of Equations Related to the First Law of Thermodynamics." *Journal of Chemical Education 87* (7): pp. 750–5. doi: 10.1021/ed1001625.

Holmes, N. G., C. Wieman, and D. A. Bonn. 2015. "Teaching Critical Thinking." *Proceedings of the National Academies of Sciences 112* (36): pp. 11199–204. doi: 10.1073/pnas.1505329112.

Huber, M. T. 2008. "The Promise of Faculty Inquiry for Teaching and Learning Basic Skills." *Strengthening Pre-collegiate Education in Community Colleges (SPECC)*. Stanford: The Carnegie Foundation for the Advancement of Teaching.

Kandel, M. 1997. "Miles Pickering—A Bibliography: A Chemistry Educator's Legacy." *Journal of College Science Teaching* 27 (3): pp. 174–8.

Marcus, A. I. 2005. *Engineering in a Land-grant Context: The Past, Present, and Future of an Idea*. West Lafayette: Purdue University Press.

O'Connor, J. J., and E. F. Robertson. 2000. "E. F. Hans Freudenthal," MacTutor History of Mathematics Archive. Accessed August 29, 2017. http://www-history.mcs.st-andrews.ac.uk/Biographies/Freudenthal.html.

Shulman, L. 1993. "Teaching as Community Property: Putting an End to Pedagogical Solitude." *Change* 25 (6): pp. 6–7.

Shulman, L. 1999. "Taking Learning Seriously." *Change* 31 (4): pp. 10–17.

Smith, M. K, W. B. Wood, W. K. Adams, C. Wieman, J. K. Knight, N. Guild, and T. T. Su. 2009. "Why Peer Discussion Improves Student Performance on In-Class Concept Questions." *Science 323* (5910): pp. 122–4. doi: 10.1126/science.1165919.

Tinberg, H., D. K. Duffy, and J. Mino. 2007. "The Scholarship of Teaching and Learning at the Two-Year College: Promise and Peril." *Change* 39 (4): pp. 26–33. doi:10.3200/CHNG.39.4.26-35.

Wieman, C. 2007. "Why Not Try a Scientific Approach to Science Education?" *Change* 39 (5): pp. 9–15.

Wieman, C. 2017. *Improving How Universities Teach Science: Lessons from the Science Education Initiative*. Boston: Harvard University Press.

CHAPTER 1

Understanding the Scholarship of Teaching and Learning (SoTL)

Introduction

All faculty members in higher education possess disciplinary knowledge, habits of mind, and skills that they have learned in their graduate studies and research. In the fields of science, technology, engineering, and mathematics (STEM), these habits of mind include taking a scientific approach to their work and a desire to improve understanding of basic principles in their field. Those same traits are available to apply to their teaching and their students' learning.

All instructors want their students to learn. Most will attempt to make some adjustments to their teaching if their students do not learn. Afterwards, they will have a gut feeling about whether the result was better. But the key to improving teaching and learning is to determine if the changes did make a measurable difference. For those not trained in **discipline-based education research** (DBER), **the scholarship of teaching and learning** (SoTL) offers a way to approach teaching and learning scientifically. SoTL—with its emphasis on identifying and defining a problem, systematically gathering evidence, and drawing conclusions from that evidence—should be attractive to scientists, engineers, and mathematicians as a logical approach to understanding and improving student learning.

The Scholarship of Teaching and Learning, Jacqueline M. Dewar, Curtis D. Bennett, and Matthew A. Fisher.
© Oxford University Press, 2018. Published 2018 by Oxford University Press

The Origins of SoTL in Higher Education

The history of SoTL may be unfamiliar to many faculty members in STEM fields. Many scholarly activities are labeled SoTL, some appropriately and others not. We begin with an account of the origins of the scholarship of teaching and learning movement and the early efforts to forge connections between SoTL and academic disciplines in the United States. We then explore the distinctions among SoTL, good teaching, scholarly teaching, and, to the extent possible, research on undergraduate teaching and learning in STEM fields. Since the last of these varies by discipline, we provide general observations on DBER and then, using the discipline of mathematics as a case study, we explore in more depth the distinction between SoTL and DBER. Finally, the chapter addresses the issue of evaluating and valuing this work, and closes with a discussion of implications for junior faculty who wish to engage in SoTL.

While SoTL now has practitioners around the world, its early development took place in the United States and that is what we describe in this chapter. In 1990, Ernest Boyer, President of the Carnegie Foundation for the Advancement of Teaching, introduced the expression "scholarship of teaching" into the vocabulary of higher education. His book, *Scholarship Reconsidered* (Boyer 1990), called for colleges and universities to embrace a broader vision of scholarship in order to tap the full range of faculty talents across their entire careers and to foster vital connections between academic institutions and their surrounding communities. Boyer argued for the recognition of **four types of scholarship**: **discovery, application, integration,** and **teaching**. The scholarship of discovery refers to what is traditionally called research in most disciplines. The scholarship of application, now sometimes called **scholarship of engagement**, refers to applying knowledge to consequential problems, often conducted with and for community partners. The scholarship of integration makes connections between disciplines, for example, by doing work at the nexus of two disciplines, applying theories used in one discipline to questions that arise in a different field, or fitting existing research into a larger framework. More and more, this kind of interdisciplinary work is being recognized as essential to solving complex real-world problems. Collaborative efforts by public health, biology, and mathematics faculty members to understand the ecology of homelessness provide an example of the scholarship of integration. Boyer's scholarships may have different interpretations depending on the discipline and type of institution. For example, for engineers, consulting work is often considered scholarship of application. Boyer's description of the fourth form, the scholarship of teaching, contained many of the characteristics of what is now called the "scholarship of teaching and learning," or "SoTL," but it failed to include peer review and making results public. Later on, for many in the SoTL movement, a fully developed definition of SoTL included these elements (Hutchings and Shulman 1999; Richlin 2001, 2003; Smith 2001). So, both the name and the concept of this form of scholarship have evolved.

As President of the Carnegie Foundation for the Advancement of Teaching, Boyer was able to bring national and international attention to SoTL, but others had discussed similar ideas even earlier. For example, Cross (1986) had argued that faculty should undertake research on teaching and learning in their classrooms in order to discover more effective teaching methods and to establish a body of knowledge about college teaching that would maximize learning. More than twenty years later, calls to do this continued (Schmidt 2008).

After Boyer, the next Carnegie President, Lee Shulman, and Vice President Pat Hutchings stated that the scholarship of teaching is the integration of the experience of teaching with the scholarship of research (Hutchings and Shulman 1999). More recently, Hutchings and Carnegie Senior Scholar Mary Huber (Huber and Hutchings 2005) have "come to embrace a capacious view of the topic, wanting to draw this movement in the broadest possible terms" (4). In their view, SoTL could range from modest investigations that document the teaching and learning in a single classroom to broad studies with elaborate research designs.

In this book we define SoTL as:

> the intellectual work that faculty members do when they use their disciplinary knowledge to investigate a question about their students' learning (and their teaching), gather evidence in a systematic way, submit their findings to peer review, and make them public for others to build upon.

This definition emphasizes the investigation into student learning over teaching practice, as that shift was essential to the transformation of Boyer's scholarship of teaching into Carnegie's scholarship of teaching and learning. As Hutchings and Shulman (1999) wrote: "Indeed, our guidelines for the Carnegie Scholars Program call for projects that investigate not only teacher practice but the character and depth of student learning that results (or does not) from that practice" (13). In addition, going public through a process that involves peer review is an integral component of this definition. This is necessary to fully realize Lee Shulman's idea of teaching as community property (Shulman 2004).

What Laurie Richlin wrote in 2001 remains true today: "Although a decade has passed since the idea of a scholarship of teaching entered the lexicon of

American higher education, the concept remains intertwined with the activities of scholarly teaching. Only by separating the different activities and focusing on the scholarly process can we give each the honor and rewards it deserves" (2001, 87). We elaborate on this entanglement next.

Distinguishing SoTL from "Good" and "Scholarly" Teaching

One of the sources of confusion about SoTL and its value to higher education is the lack of clear distinctions among three overlapping concepts: **good teaching**, **scholarly teaching**, and **the scholarship of teaching and learning**. While these terms can have many interpretations, we rely on Smith (2001), who wrote that good (or better) teaching is defined and measured by the quality of student learning. The overwhelming success of Jaime Escalante's students on the 1982 Advanced Placement Calculus exam provided clear evidence of "good" teaching (Mathews 1989). Scholarly teaching refers to something else. Scholarly teachers and their teaching must be informed not only by the latest developments in their disciplinary fields but also by research about instructional design and how students learn, either in general or in a discipline-specific way. Based on this research, scholarly teachers make choices about instruction and assessment. Deciding to use interactive engagement teaching methods so as to promote better conceptual understanding in an elementary physics course (Hake 2002) is an example of "scholarly" teaching. When instructors do more than simply draw on research to become scholarly teachers, and also seek to contribute to the knowledge base by carrying out research on teaching and learning, then they are engaging in the scholarship of teaching and learning. This discussion represents but one way to parse these terms and does not even address another closely related term, **reflective teaching**. This refers to teachers who engage in the practice of self-observation and self-evaluation.

For most STEM faculty, doing SoTL will involve aspects of discovery, application, and integration (Boyer 1990). It will be undertaken to improve practice, within or beyond the investigator's classroom, or both. But to be SoTL, the inquiry must also satisfy the three additional features of being public, open to critique and evaluation by peers, and in a form that others can build on (Hutchings and Shulman 1999). "Going public" is not an end in itself, nor the primary motivation for a SoTL project, but it is an essential part of being open to critique and something that others can build on.

Forging Disciplinary Connections for SoTL

In 1998, Carnegie initiated new programs to promote SoTL under the name Carnegie Academy for the Scholarship of Teaching and Learning (CASTL). The first was the CASTL Scholars Program, which selected 158 post-secondary faculty members, many distinguished researchers in their disciplines, to populate six cohorts of CASTL scholars over nine years. They worked on individual scholarship of teaching and learning projects and many went on to become leaders in the SoTL movement. Among these scholars, 48 were chosen from the fields of biology, chemistry, computer science, earth science, geology, medical and allied health sciences, engineering, mathematics and statistics, physics, or science education (see Appendix I for a list of scholars from STEM and related fields).

Carnegie also initiated the Scholarly and Professional Societies Program to encourage recognition of SoTL by the disciplines. The disciplines of biology, mathematics, physics, and several of the social sciences (economics, psychology, sociology, and political science) were represented among the 28 participating societies (see Appendix II for a list of participants). Five health science organizations, including one in England, were also involved. During the same period several Carnegie publications and speeches approached SoTL via the disciplines (Huber and Morreale 2002; Shulman 2005).

Halfway through the decade-long CASTL Scholars Program, Carnegie began two new programs aimed at building support for SoTL at the institutional level, the CASTL Institutional Leadership Clusters (2003–06) and the Carnegie Institutional Leadership and Affiliates Program (2006–09). Over 150 colleges and universities from Australia, Canada, the United Kingdom, and the United States (US) participated in one or both programs. A more detailed discussion of the succession of programs that ran under Carnegie's long-term initiative to promote Boyer's vision of teaching as scholarly work can be found in *Scholarship of Teaching and Learning Reconsidered* (Hutchings et al. 2011). By the time the CASTL initiative came to a formal close in 2009, SoTL had become an increasingly international movement. The International Society for the Scholarship of Teaching and Learning, founded in 2004, is now the leading international society promoting and supporting the scholarship of teaching and learning. Meanwhile, interest in improving undergraduate education, especially in STEM fields led groups in the US, such as the National Academies, the Association of American Colleges and Universities, and the National Center for Science and Civic Engagement, to shared agendas that draw on or contribute to the advancement of the scholarship of teaching and learning.

Discipline-Based Education Research (DBER)

In many countries around the world, recent reports and initiatives have called attention to the need to improve school and university-level STEM education in order to prepare a diverse technical workforce and educate a science-literate citizenry. They also advocated for more research, both theoretical and applied, on STEM teaching and learning (Caprile et al. 2015; Office of the Chief Scientist 2013). Examples of such reports generated in the US include the National Research Council (2012, 2015) and the President's Council of Advisors on Science and Technology (2012). Teachers of pre-college and undergraduate science have been exhorted to apply the results of this research to their teaching practice and often to engage in studying their own practice. For example, in its 2010 position paper on the role of research on science teaching and learning, the National Science Teachers Association (NSTA) urged: "NSTA encourages ALL participants in science education, including K–16 teachers of science and administrators, to recognize the importance of research and assume active roles in research practices" (National Science Teachers Association 2010, 1).

In the US, in most STEM disciplines, research on teaching and learning predated the origin of SoTL. For example, physics education research can trace its roots back more than four decades to the 1970s when the first doctoral degrees in physics for research on learning and teaching of physics were awarded (National Research Council 2013, 59). In chemistry, the *Journal of Chemistry Education* has provided a forum for sharing ideas and research on teaching for over 90 years. According to Felder and Hadgraft (2013), in the 1980s engineering education shifted from reliance on student satisfaction surveys and instructor impressions to studies involving statistical comparisons between experimental and control groups and then, in the twenty-first century, began to adopt social science methodologies. Wankat et al. (2002) described how changes in engineering accreditation (ABET 1997) that emphasized the formulation and assessment of learning outcomes gave a boost to engineering education research and to SoTL in engineering. Biology education research emerged more recently than similar efforts in physics, chemistry, or engineering (Singer et al. 2013).

The contributions of discipline-based education researchers in biology, physics and astronomy, chemistry, engineering, and the geosciences were the focus of a 2012 report, *Discipline-Based Education Research: Understanding and Improving Learning in Undergraduate Science and Engineering*, by the National Research Council. The report resulted from the National Science Foundation "recognizing DBER's emergence as a vital area of scholarship and its potential to improve undergraduate science and engineering education" and "[requesting] the National Research Council convene the Committee on the Status, Contributions, and Future Directions of Discipline-Based Education Research to conduct a synthesis study of DBER" (National Research Council 2012, 1). It described DBER as distinguished by an empirical approach to investigating learning and teaching that is informed by an expert understanding of disciplinary knowledge and practice. DBER is represented as overlapping with but distinct from SoTL, educational psychology research, and cognitive science. However, the report also noted that "the boundaries between SoTL and DBER are blurred and some researchers belong to both the SoTL and DBER communities" (12). We will make a similar observation when we compare SoTL with discipline-based education research in undergraduate mathematics.

SoTL and DBER: A Case Study in Mathematics

Each STEM field has its own developmental story to tell concerning its discipline-based education research and its participation in the SoTL movement. We relate the details of this story for mathematics, a discipline not included in the National Research Council (2012) report. Readers from other STEM disciplines may find it instructive to compare and contrast our description of the situation in mathematics with that in their own fields.

SoTL first entered the mathematics community in the US in a visible way in 1999 with the recruitment of four professors of mathematics, including the President of the Mathematical Association of America (MAA), to the CASTL Scholars Program. Over the next six years, another eight mathematicians (including two of the authors) participated in the program.

In 2007, those two authors developed a four-hour introductory workshop on doing SoTL that has since been presented at eight national mathematics meetings to more than 200 people. In addition, SoTL paper sessions have been part of every national winter mathematics meeting since 2007. Some research topics presented in the early SoTL paper sessions, such as getting students to read the textbook, anticipated growing interest on the part of the mathematics community and later appeared as the specific focus of other contributed paper sessions. SoTL is now frequently a focus of professional development for junior faculty through the MAA's New Experiences in Teaching Project, suggesting that SoTL will have a lasting presence in the mathematical community. By 2015, SoTL had garnered enough interest and recognition within the discipline of mathematics to prompt the MAA to publish a book, *Doing the Scholarship of Teaching and Learning in Mathematics* (Dewar and Bennett 2015) in its MAA Notes series.

As the SoTL movement unfolded and called for faculty to treat teaching and learning in a scholarly fashion, mathematics already had a preexisting community of scholars doing educational research into collegiate level mathematics teaching, the Special Interest Group of the MAA on Research in Undergraduate Mathematics Education (**RUME**) (see http://sigmaa.maa.org/rume/Site/About.html). Currently there are 77 doctoral programs in mathematics education in the US, 72 of which grant a PhD (as opposed to an EdD) degree (Hsu 2013). RUME is set apart from disciplinary mathematical research by its very different questions of interest, methodologies, and epistemologies (Schoenfeld 2000), but how it differs from SoTL carried out in mathematics may not be so clear.

Mathematicians and Carnegie scholars Tom Banchoff and Anita Salem (2002) saw SoTL as potentially bridging the gap between RUME and teaching mathematicians, but we prefer a different metaphor: SoTL as part of a broadened landscape of scholarly work related to teaching. Our model, shown in Figure 1.1, places the labels "Teaching Tips," "SoTL," and "Mathematics Education Research" at the three vertices of an equilateral triangle. The result is a space that can be populated by all sorts of work, including that of mathematicians who develop new ways of using technology for teaching and learning mathematics. Our description of this shared space aligns with Huber and Hutchings' (2005) view of SoTL as increasing **the teaching commons**, "a conceptual space in which communities of educators committed to inquiry and innovation come together to exchange ideas about teaching and learning and use them to address the challenges of educating students for personal, professional, and civic life" (*x*).

In our model, teaching tips refers to a description of a teaching method or innovation that an instructor reports having tried "successfully" and that the students "liked." If the instructor begins to systematically gather evidence from students about what, if any, cognitive or affective effect the method had on their

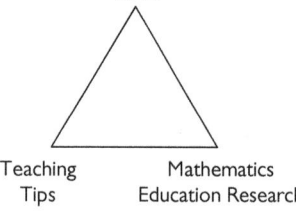

Figure 1.1 The teaching commons: a broadened landscape of scholarly work related to teaching.

learning, he or she is moving toward scholarship of teaching and learning. When this evidence is sufficient to draw conclusions, and those conclusions are situated in the literature, peer reviewed, and made public, the instructor has produced a piece of SoTL work. However, the work may differ in form and scope from that found at the third vertex of the triangle in Figure 1.1.

Mathematics education research, or RUME as we will refer to it, is more in line with Boyer's "scholarship of discovery," wherein research methodologies, theoretical frameworks, empirical studies, and reproducible results would command greater importance. This naturally influences the questions asked or considered worth asking, the methods used to investigate them, and what the community accepts as valid. The work of Schwab (1964) on the **structure of the disciplines** provides four ways of identifying and describing the differences between SoTL and RUME.

1 *Content boundaries*: RUME is focused on mathematical knowledge and mathematical learning, and is interdisciplinary in that it draws on studies in education, psychology, sociology, anthropology, and linguistics as well as its own body of literature. SoTL, while being interdisciplinary and having a growing body of literature, is a multi-disciplinary community with practitioners in disciplines ranging from the arts and humanities, to business, and to the sciences and engineering. Mathematicians in the SoTL community encounter SoTL investigations in these other disciplines, encouraging their interest in questions that cross boundaries, for example, questions about **student voice** (that is, student perspectives and insights on learning, teaching, and curriculum, and considerations of how to incorporate them into decision-making processes) or the impact of community-based-learning experiences.

2 *Skills and habits employed by practitioners*: SoTL questions virtually always come from practice. SoTL researchers encounter a teaching or learning problem they want to fix, a question they want to answer, or a phenomenon they desire to understand or describe in depth. Sometimes SoTL practitioners simply aim to show what is possible in a certain situation. RUME researchers may happen upon a question through their teaching practice, but more frequently their questions arise from theory or methodological concerns. Doctoral training in mathematics education research gives them a formal background in theoretical frameworks for mathematical knowledge and learning

that they can draw upon. In contrast, SoTL researchers may utilize the taxonomy of SoTL questions—*What works? What is? What could be?*—to refine their questions (Hutchings 2000). (The process of developing and refining SoTL research questions by using this taxonomy and disciplinary knowledge is the focus of Chapter 2.) They may call on the theoretical frameworks developed by the RUME community and must, of course, appropriately place their work in the context of previous studies in order to publish it.

3 *Modes of inquiry*: Compared to the typical SoTL practitioner, the typical RUME researcher has far more expertise and knowledge about research methodologies and ways of communicating methodological concerns. Still, the SoTL researcher seeks to triangulate the evidence, that is, to use data from several sources, in support of a conclusion, and may draw on both quantitative and qualitative data. (Chapter 6 is intended to inform STEM instructors about possibly unfamiliar methods for gathering and analyzing evidence, such as interviews, focus groups, and think-alouds, that could prove useful in a SoTL study.)

4 *Purposes or outcomes for the disciplinary work*: The SoTL practitioner seeks to answer a question about teaching and learning that most likely arose from teaching. The work is made public to improve teaching practice and student learning elsewhere. As a research discipline of its own, RUME seeks to increase what is known and to develop and extend theories about mathematics learning at its deepest level. RUME also seeks to improve teaching practice and student learning.

Our elaboration of these differences specifically for RUME versus SoTL in mathematics in the US should help those in other STEM fields think about whether and how SoTL differs from the education research done in their discipline. Depending on the discipline or the country, they may be able to find similar comparisons and contrasts.

The *intended audience* may be yet another way to differentiate SoTL from DBER. Again, we provide details from the discipline of mathematics. Since its inception, SoTL has been aimed at instructors in all fields of higher education, with the intent to improve teaching and learning and to provide a basis for others to build upon. As a research field of its own, other RUME researchers naturally constitute the primary audience for original RUME work. But, the line of demarcation can be blurry. Two members of the RUME community,

Marilyn Carlson and Chris Rasmussen (2008), edited a volume in the MAA Notes series, *Making the Connection: Research and Teaching in Undergraduate Mathematics Education*, which was "intended for mathematicians and mathematics instructors who want to enhance the learning and achievement of students in their undergraduate mathematics courses" (*vii*). Of course, almost no form of scholarship fits neatly into any one camp. SoTL work has been presented at RUME conferences and RUME researchers have presented work in SoTL paper sessions at the national mathematics meetings. Thus the current situation in the mathematical community seems to support Huber and Hutchings (2005) in their claim that the teaching commons can be a big tent whose purpose is to improve teaching and learning as a whole.

Evaluating SoTL

Each discipline, institution, and department has a perspective on SoTL (and, for that matter, on discipline-based education research) and its appropriate place within the workload of a tenured or tenure-track faculty member. Physics was the first of the STEM fields to make an official statement by its leading professional society about the value of physics education research and how it should be evaluated. The Council of the American Physical Society adopted a statement on May 21, 1999, supporting the acceptance of physics education research by physics departments that reads in part, "Physics education research can and should be subject to the same criteria for evaluation (papers published, grants, etc.) as research in other fields of physics" (American Physical Society 1999).

Chemistry too has been very supportive of SoTL. The *Statement on Scholarship* of the American Chemical Society, approved in 2010, clearly embraced the broader view of scholarship promoted by Boyer: "Rigorous scholarship in discovery, integration, application, and the study of teaching and learning is needed to foster the innovations that will ensure our economic health in a global economy" (1). It further signaled its support by noting that while SoTL may be the least understood and recognized of all the forms of scholarship, it "has the potential to transform chemical education. It must be encouraged and its role in preparing scientists for the new millennium must be recognized" (American Chemical Society 2010, 2).

In many departments, SoTL-type work is not viewed as disciplinary research unless the faculty member's position was designated as a pedagogical researcher.

So the question naturally arises: By what standards should SoTL work be evaluated?

The question is not new. Soon after the publication of *Scholarship Reconsidered* (Boyer 1990), it became clear that an essential piece of promoting a broader definition of scholarship was missing. The effort to broaden the meaning of scholarship simply could not succeed until institutions had clear standards for evaluating this wider range of scholarly work. Faculty and administrators accord full academic value only to work they can confidently judge, through a process usually involving peer review. That prompted senior staff members at the Carnegie Foundation to undertake a new project. They collected and examined hundreds of documents (tenure and promotions guidelines, federal grant and private foundation funding criteria, academic journal submission guidelines, etc.) in an attempt to distill a common set of standards. The result, *Scholarship Assessed* (Glassick et al. 1997), became another foundational SoTL publication. According to Glassick et al., the most remarkable feature of this collection of guidelines was how much they shared common elements. Their synthesis of these materials provided a clear response to the problem of how to judge SoTL: SoTL, or any of the other nontraditional forms of scholarship, is to be judged by the same criteria as the traditional scholarship of discovery. Their book delineated six standards for assessing any scholarly work, and provided questions to flesh out the meaning of each standard:

1 Clear goals—Does the scholar state the basic purposes of his or her work clearly? Does the scholar define objectives that are realistic and achievable? Does the scholar identify important questions in the field?
2 Adequate preparation—Does the scholar show an understanding of existing scholarship in the field? Does the scholar bring the necessary skills to his or her work? Does the scholar bring together the resources necessary to move the project forward?
3 Appropriate methods—Does the scholar use methods appropriate to the goals? Does the scholar apply effectively the methods selected? Does the scholar modify procedures in response to changing circumstances?
4 Significant results—Does the scholar achieve the goals? Does the scholar's work add consequentially to the field? Does the scholar's work open additional areas for further exploration?
5 Effective presentation—Does the scholar use a suitable style and effective organization to present his or her work? Does the scholar

use appropriate forums for communicating work to its intended audiences? Does the scholar present his or her message with clarity and integrity?

6 Reflective critique—Does the scholar critically evaluate his or her own work? Does the scholar bring an appropriate breadth of evidence to his or her critique? Does the scholar use evaluation to improve the quality of future work? (Glassick et al. 1997, 36)

This widely cited work has, in theory, answered the question of how SoTL work can be evaluated.

Valuing SoTL: Special Considerations for Junior Faculty Members

Still, whether and, if so, how much individual colleagues, departments, and institutions count SoTL as research in tenure and promotion decisions varies widely. At some institutions SoTL will be relegated to the teaching section of a tenure dossier where it can perform a useful function as evidence of teaching effectiveness and of going beyond scholarly teaching. Others may consider a SoTL article as little more than an addendum to a required level of traditional disciplinary research, especially when published in an interdisciplinary SoTL journal rather than in a disciplinary or STEM education publication. Downgrading based on publication venue is a problem, particularly for pre-tenure faculty members. As the National Research Council (2012) noted: "Discipline-based education researchers might encounter close scrutiny regarding the prestige of their field's journals. Faculty who are not yet tenured may question the merit of submitting manuscripts to journals with impact factors significantly lower than those in which their disciplinary peers are publishing" (38). On the other hand, the decision at *Science*, a scientific research journal with an impact factor greater than 30, to publish papers on education was seen as a significant advancement for all DBER fields (National Research Council 2012). Dewar and Bennett (2010) described the lack of SoTL publishing opportunities in mathematics journals—a problem that has continued.

Some institutions have fully embraced SoTL's inclusion as research (Gieger et al. 2015; Kaus 2015; Tate 2010), while others have tried and run into difficulty (Schweitzer 2000). William Boyd (2013), a professor of geography at Southern Cross University in Australia, represents an interesting case study in both the promise and applicability of Boyer's model for scholarship. Boyd described how the use of Boyer (1990) and Glassick et al. (1997) as a unifying frame for his interdisciplinary work that draws on the geosciences and the humanities led to his success in four areas of academic life: professional promotion, peer mentoring, developing cross-disciplinary work, and curriculum development. Huber (2004) examined the careers of four academics at research-intensive institutions who chose to make SoTL a significant part of their research. Two STEM fields were represented: chemistry and mechanical engineering. All four achieved successful outcomes as measured by obtaining tenure and promotion, but not without facing challenges and difficulties. Since the conversation initiated by Boyer in 1990 to move beyond a narrow definition of research has yet to come to a full and satisfying conclusion (O'Meara and Rice 2005), pre-tenure faculty members should seek guidance and clarity from their departments and institutions on how SoTL work will be valued.

The growing number of teaching-focused faculty positions around the world is another development. These positions come with widely varying expectations for research and possibilities for periodic renewal or permanence. Annetta Tsang (2010) described her shift from a traditional academic position in her field of dentistry and oral health to a teaching-focused one at the University of Queensland. In the process she gained many insights into the role SoTL can play in such a position. Her advice for anyone interested in establishing a SoTL academic pathway (Tsang 2010) ranged from topics already considered in this chapter (such as distinguishing SoTL from scholarly teaching) to many that we will discuss in Chapter 8 (such as finding collaborators and allocating time for scholarly writing).

Throughout this book we periodically pause the narrative to insert questions that prompt readers to reflect, plan, or act. Each set of questions to consider is connected to a common theme, which we identify. They are clearly marked as a place to "Pause" and the first of these follows.

Pause: Before Starting an Investigation—Questions to Consider

- In terms of my professional life, what is my goal in undertaking a SoTL investigation?
- What is the relationship between SoTL and DBER in my discipline?
- How is SoTL valued in my department and institution?
- What role can SoTL play in my tenure, promotion, or merit application?

REFERENCES

Accreditation Board for Engineering and Technology. 1997. *Engineering Criteria 2000.* Baltimore: ABET.

American Chemical Society. 2010. "Statement on Scholarship." Accessed May 26, 2017. https://www.acs.org/content/dam/acsorg/about/governance/committees/education/statement-on-scholarship.pdf.

American Physical Society. 1999. "Research in Physics Education." Accessed May 26, 2017. http://www.aps.org/policy/statements/99_2.cfm.

Banchoff, T., and A. Salem. 2002. "Bridging the Divide: Research versus Practice in Current Mathematics Teaching and Learning." In *Disciplinary Styles in the Scholarship of Teaching and Learning: Exploring Common Ground*, edited by M. T. Huber and S. P. Morreale, pp. 181–96. Washington: American Association for Higher Education.

Boyd, W. 2013. "Does Boyer's Integrated Scholarships Model Work on the Ground? An Adaption of Boyer's Model for Scholarly Professional Development." *International Journal for the Scholarship of Teaching and Learning* 7(2). doi: 10.20429/ijsotl.2013.070225.

Boyer, E. L. 1990. *Scholarship Reconsidered: Priorities of the Professoriate.* San Francisco: Jossey-Bass.

Caprile, M., R. Palmén, P. Sanz, and G. Dente, G. 2015. *Encouraging STEM Studies Labour Market Situation and Comparison of Practices Targeted at Young People in Different Member States.* European Parliament: Policy Department A: Economic and Scientific Policy. Accessed August 30, 2017. http://www.europarl.europa.eu/RegData/etudes/STUD/2015/542199/IPOL_STU(2015)542199_EN.pdf.

Carlson, M., and C. Rasmussen, editors. 2008. *Making the Connection: Research and Teaching in Undergraduate Mathematics Education* (MAA Notes, #73). Washington: Mathematical Association of America.

Cross, K. 1986. "A Proposal to Improve Teaching or What 'Taking Teaching Seriously' Should Mean." *AAHE Bulletin 39* (1): pp. 9–14.

Dewar, J., and C. Bennett. 2010. "Situating SoTL Within the Disciplines: Mathematics in the United States as a Case Study." *International Journal of the Scholarship of Teaching and Learning 4* (1). doi: 10.20429/ijsotl.2010.040114.

Dewar, J., and C. Bennett, editors. 2015. *Doing the Scholarship of Teaching and Learning in Mathematics.* Washington: Mathematical Association of America.

Felder, R., and R. Hadgraft. 2013. "Educational Practice and Educational Research in Engineering: Partners, Antagonists, or Ships Passing in the Night?" *Journal of Engineering Education 102* (3): 339–45. doi: 10.1002/jee.20015.

Gieger, L., J. Nardo, K. Schmeichel, and L. Zinner. 2015. "A Quantitative and Qualitative Comparison of Homework Structures in a Multivariable Calculus Class." In *Doing the Scholarship of Teaching and Learning in Mathematics*, edited by J. Dewar and C. Bennett, pp. 67–76. Washington: Mathematical Association of America.

Glassick, C., M. Huber, and G. Maeroff. 1997. *Scholarship Assessed: Evaluation of the Professoriate.* San Francisco: Jossey-Bass.

Hake, R. 2002. "Lessons from the Physics Education Reform Effort." *Conservation Ecology 5* (2), Article 28. Accessed September 1, 2017. http://www.consecol.org/vol5/iss2/art28.

Hsu, E. 2013. *Spreadsheet of North American Doctoral Programs in Math Education*. Accessed August 29, 2017. http://sigmaa.maa.org/rume/phd.html.

Huber, M. T., and P. Hutchings. 2005. *The Advancement of Learning: Building the Teaching Commons*. San Francisco: Jossey-Bass.

Huber, M. T. 2004. *Balancing Acts: The Scholarship of Teaching and Learning in Academic Careers*. Sterling: Stylus Publishing.

Huber, M. T., and S. P. Morreale, editors. 2002. *Disciplinary Styles in the Scholarship of Teaching and Learning: Exploring Common Ground*. Washington: American Association for Higher Education.

Hutchings, P., editor. 2000. *Opening Lines: Approaches to the Scholarship of Teaching and Learning*. Palo Alto: The Carnegie Foundation for the Advancement of Teaching.

Hutchings, P., M. Huber, and A. Ciccone. 2011. *The Scholarship of Teaching and Learning Reconsidered*. San Francisco: Jossey-Bass.

Hutchings, P., and L. Shulman. 1999. "The Scholarship of Teaching: New Elaborations, New Developments." *Change 31* (5): pp. 10–15.

Kaus, C. 2015. "Using SoTL to Assess the Outcomes of Teaching Statistics through Civic Engagement." In *Doing the Scholarship of Teaching and Learning in Mathematics*, edited by J. Dewar and C. Bennett, pp. 99–106. Washington: Mathematical Association of America.

Mathews, J. 1989. *Escalante: The Best Teacher in America*. New York: H. Holt.

National Research Council. 2012. *Discipline-Based Education Research: Understanding and Improving Learning in Undergraduate Science and Engineering*. Washington, DC: The National Academies Press. doi: 10.17226/13362.

National Research Council. 2013. *Adapting to a Changing World—Challenges and Opportunities in Undergraduate Physics Education*. Washington, DC: The National Academies Press. doi: 10.17226/18312.

National Research Council. 2015. *Reaching Students: What Research Says About Effective Instruction in Undergraduate Science and Engineering*. Washington: The National Academies Press. doi: 10.17226/18687.

National Science Teachers Association. 2010. *NSTA Position Statement: The Role of Research on Science Teaching and Learning*. Accessed August 29, 2017. http://www.nsta.org/about/positions/research.aspx.

Office of the Chief Scientist. 2013. *Science, Technology, Engineering and Mathematics in the National Interest: A Strategic Approach*. Canberra: Australian Government, Canberra. Accessed August 29, 2017. http://www.chiefscientist.gov.au/wp-content/uploads/STEMstrategy290713FINALweb.pdf.

O'Meara, K., and E. Rice, editors. 2005. *Faculty Priorities Reconsidered: Rewarding Multiple Forms of Scholarship*. San Francisco: Jossey-Bass.

President's Council of Advisors on Science and Technology. 2012. *Engage to Excel: Producing One Million Additional College Graduates with Degrees in Science, Technology, Engineering, and Mathematics*. Washington: Executive Office of the President, President's Council of Advisors on Science and Technology.

Richlin, L. 2001. "Scholarly Teaching and the Scholarship of Teaching." In *Scholarship Revisited: Perspectives on Scholarship of Teaching and Learning*, New Directions for Teaching and Learning: No. 86, edited by C. Kreber, pp. 57–68. San Francisco: Jossey-Bass.

Richlin, L. 2003. "Understanding, Promoting and Operationalizing the Scholarship of Teaching and Learning: A Message from the Editor." *Journal on Excellence in College Teaching 14* (2/3): pp. 1–4.

Schmidt, P. 2008. "Harvard's Derek Bok: Professors, Study Thine Own Teaching." *Chronicle of Higher Education*, October 13. Accessed August 30, 2017. http://www.chronicle.com/article/Harvards-Derek-Bok-/1239.

Schoenfeld, A. H. 2000. "Purposes and Methods of Research in Mathematics Education." *Notices of the American Mathematical Society 47* (6): pp. 641–9.

Schwab, J. 1964. "Structure of the Disciplines." In *The Structure of Knowledge and the Curriculum*, edited by G. W. Ford and L. Pugno, pp. 6–30. Skokie: Rand McNally.

Schweitzer L. 2000. "Adoption and Failure of the 'Boyer Model' at the University of Louisville." *Academic Medicine 75* (9): pp. 925–9.

Shulman, L. 2004. *Teaching as Community Property: Essays on Higher Education*. San Francisco: Jossey-Bass.

Shulman, L. 2005. "Signature Pedagogies in the Professions." *Daedalus 134* (3): pp. 52–9. doi: 10.1162/0011526054622015.

Singer, S., N. Nielsen, and H. Schweingruber. 2013. "Biology Education Research: Lessons and Future Directions." *CBE Life Sciences Education 12* (2): pp. 129–32. doi: 10.1187/cbe.13-03-0058.

Smith, R. 2001. "Formative Evaluation and the Scholarship of Teaching and Learning." In *Fresh Approaches to the Evaluation of Teaching*, New Directions for Teaching and Learning: No. 88, edited by C. Knapper and P. Cranton, pp. 51–62. San Francisco: Jossey-Bass.

Tate, T. 2010. "Trailblazing in Academics: Novel Policies that Reward Faculty for Work Beyond Traditional Research and Publications Garner National Attention as WCU Embraces the Boyer Model." *The Magazine of Western Carolina University 14* (3): pp. 20–5.

Tsang, A. 2010. "Pitfalls to Avoid in Establishing A SoTL Academic Pathway: An Early Career Perspective." *International Journal for the Scholarship of Teaching and Learning 4* (2). doi: 10.20429/ijsotl.2010.040219.

Wankat, P., R. Smith., K. Felder., and F. Oreovicz. 2002. "The Scholarship of Teaching and Learning in Engineering." In *Disciplinary Styles in the Scholarship of Teaching and Learning: Exploring Common Ground*, edited by M. Huber and S. Morrealle, pp. 217–37. Washington: AAHE/Carnegie Foundation for the Advancement of Teaching.

CHAPTER 2

Developing a Researchable Question

Introduction

In one of the formative articles of the scholarship of teaching and learning (SoTL) movement in the United States (US), Randy Bass (1999) discussed the divergent reactions that "teaching problems" and "research problems" typically elicit from faculty members, with the latter engendering far more positive interest and reaction. He posited that one of the tenets of SoTL is that a teaching problem should be viewed as an invitation to a scholarly investigation, similar to how most faculty members view a research problem.

Teaching problems typically present themselves as difficulties or frustrations we encounter in our teaching. For example, *my students aren't as prepared for class as I would like them to be, they don't do the required reading*, and *they only start working on an assignment the night before it is due* are all problems that will sound familiar to many instructors. Our goal in this chapter is to show how to take a teaching problem and transform it into a researchable question. Attempting to answer such a question is very likely to involve gathering evidence from students. If so, it is considered **human subjects research** and should be cleared through the **institutional review board** on campus. The approval process for human subjects research is discussed in Chapter 3 (Human Subjects Considerations).

A Taxonomy of SoTL Questions

In converting problems to researchable questions, it helps to understand the types of questions that can be researched. Carnegie Vice President Pat Hutchings (2000) identified the following **taxonomy of SoTL questions** based upon seeing the work of many scholars in the Carnegie Academy for the Scholarship of Teaching and Learning (CASTL) Scholars Program.

> *What works?*—This question asks whether a particular teaching method, assignment, or approach is effective.
> *What is?*—This question seeks detailed information about what students are really doing, thinking, or feeling in a particular situation.
> *What could be?*—This question examines a case where something really interesting happened or a situation involving something the instructor is passionate about or has an opportunity to do. This question type aims to show what is possible, so it is sometimes referred to as *a vision of the possible.*

Hutchings (2000) also acknowledged *theory-building* as a fourth but less common type of SoTL work. Theory-building is a far more common purpose for discipline-based education research (DBER) than for SoTL.

When investigating a proposed solution to a problem, it is natural to begin with a *What works?* question. For example, for a problem with students not doing the reading for a class, we might attempt to solve the problem by giving reading quizzes as suggested by Felder and Brent (1996). This intervention might lead to the initial *What works?* question: "Does giving reading quizzes make students read the text more carefully before class?" Physics professors Henderson and Rosenthal (2006) found reading quizzes to have shortcomings and ended up altering their class by requiring students to submit questions on the assigned reading before class. This led to a SoTL study investigating the *What works?* question: "Does having students generate questions on what they read contribute to student learning in an introductory physics class?"

The goal of a *What works?* question is to investigate whether a particular teaching technique or assignment is effective and leads to better student outcomes. Here are other examples of *What works?* questions, including some that have been investigated and published (citations are provided for the published studies):

- Does infusing biology applications into a calculus-for-the-life-sciences course improve biology majors' attitudes about learning calculus?

- Does using interactive technology in an introduction to engineering course improve student performance and retention?
- Does the use of robots as a context-based learning tool in an introductory computer science course improve student learning? (Rolka and Remshagen 2015)
- What is the effect of three different types of recitation sessions in a college calculus course on raising final examination scores, lowering DFW rates, and increasing one-year retention rates for STEM majors? (Watt et al. 2014)

What is? questions often arise when we want to know what students are doing or thinking or we want to understand why something is happening. *What is?* studies may lead to powerful results that open a window to the student mind. Here are some examples:

- What do physics students think the purpose of taking lab classes is?
- What kind of difficulties do multi-step problems pose for students?
- Why don't C, D, and F students review their test papers and learn from their mistakes?
- How are various representations and prototypes used in designing a lab-on-a-chip device in a systems biology laboratory? (Aurigemma et al. 2013).
- What is it that students do when they solve statics problems? (Litzinger et al. 2010).

Some SoTL studies involve an interplay between *What is?* and *What works?* questions. For example, in a study of his introductory statistics course for engineering majors, Bruff (2015) first had to answer the *What is?* question: "What are students able to learn by reading their textbooks before class?" before he could begin to investigate the *What works?* question that he had originally posed: "What kinds of pre-class reading assignments, including questions about the reading, might help students to learn more from reading their textbooks before class?"

Similarly, Offerdahl and Montplaisir (2014) first investigated the *What is?* question: "What types of questions do students in an upper-level, large-lecture biochemistry course generate from their reading assignments?" Then they used this data to answer a *What works?* question: "Could student-generated reading questions be used effectively as formative feedback?"

The third type of question, the *What could be?* question, might arise when experimenting with a new type of course or classroom activity or when something unusual or good happens in a class. Examples of these types of questions include:

- Each semester the ability of a few students to solve problems improves dramatically during a semester-long course. Can I document the progression of such a student?
- Is it possible for students in an introductory statistics course that involves projects using real world data to demonstrate competence in authentic data analysis? (Holcomb 2015)
- What happens if students in a science course for non-science majors are given the opportunity to use what they learn in the college classroom to develop curricula and then teach it in an elementary school? (Elmendorf 2006)

Not surprisingly, *What could be?* investigations may involve other types of questions as well. For example, the opportunity afforded by a grant to add group projects on community issues to a quantitative literacy course (Dewar et al. 2011) led to a *What could be?* question:

- What happens if group projects on community issues are added to a quantitative literacy course?

Consideration of this question prompted more questions:

- Will students learn as much or more?
- Will their attitudes toward mathematics change?
- What else might they gain?

Several of these subordinate questions were then re-framed as *What works?* questions:

- Will students who do group projects on community issues in a quantitative literacy course learn more mathematics than students taught in the regular lecture-discussion format?
- Will they leave the course with more positive attitudes toward mathematics and its usefulness?

As we have noted several times, SoTL projects can involve several different types of questions simultaneously. The types of questions asked will have an influence on the design of the study and the kind of data—quantitative, qualitative, or both—that are collected. We will discuss the connection between the taxonomy question types and the research design (see Chapter 3) and the evidence we might plan to gather in a study (see Chapters 4–6). For the remainder of this chapter we focus on how to transform a teaching problem into a researchable question.

Generating Questions from Problems

To take a teaching problem and turn it into something to be investigated we view the problem as something that we can understand or solve. Thus, we might want to (1) test possible solutions to the problem, (2) discover information that might help solve the problem, or (3) simply understand the problem better. If we are already trying something that we think improves student learning, perhaps we simply want to find out if it is working. While this is often where SoTL projects start, many interesting projects evolve into studies that try to increase the investigator's understanding of student thinking.

SoTL questions can also be rooted in or informed by assessment efforts at a faculty member's institution. While **assessment** and SoTL share some features, such as a focus on student learning and an emphasis placed on gathering evidence, there are some important differences. SoTL is rooted in the desire of faculty to understand student learning and factors that affect it, as well as to develop pedagogical approaches that improve learning. Assessment is rooted in a concern by institutions to have evidence of effectiveness for programs or departments. A fuller discussion of the relationship between SoTL and assessment is beyond the scope of this book; we point interested faculty to the work of Hutchings et al. (2011).

As in traditional scholarship, we try to discover what information would be most useful for moving forward. Thus, we brainstorm what we would like to know, and we make conjectures that we might test. In our SoTL workshops, once participants have done some brainstorming, we ask them to try to generate both a *What works?* and a *What is?* question related to their teaching problem.

> ### Pause: Brainstorming—Questions to Consider
> - What is my teaching problem or question?
> - What knowledge would help me address this problem or question?
> - What would I like to know about this problem or why it occurs?
>
> *continued*

- Am I doing something to address the problem and would like to show that it is working?
- Is there something I would like students to do differently?
- Why do students behave in ways that lead to the problem?
- What is it I want to accomplish in researching this problem?
- What are *What works?* and *What is?* questions that arise from my teaching problem?

Narrowing Questions

Beginning SoTL scholars tend to generate very broad initial questions. For example: "Is it beneficial to use R (a free, open-source statistics computer language) in a mathematical statistics course?" or: "What effect does problem-based learning have on chemistry students?" In order for questions like these to be researchable, they must be narrowed in focus. Some ways to do this include: clarifying the meaning of terms in the question, restricting the scope of the investigation, or considering how knowledge or learning may be structured in ways relevant to the proposed investigation.

Consider the question: "Will having students generate reading questions work to prepare them for class?" We start by asking ourselves whether the meaning of each term or phrase in the question is clear. For example, "work" or "work to prepare them for class" might mean one or more of the following: that students are more engaged in the classroom, that more class time can be devoted to problem solving, that students gain a better conceptual understanding of the material, or that students do better on the tests. To perform a systematic investigation of this question, we would need to decide which of these interpretations we want to examine. Similarly, with this question, we should clarify what we mean by having students generate questions. Are these questions asked in class, asked of themselves, reported to the instructor, or something else entirely?

Once the terms in a question are clarified, we can further refine the question by restricting it to a smaller set of classes, students, or types of learning. For the last of these, making use of a general learning taxonomy such as **Bloom's taxonomy** or a more discipline-specific categorization may lead us to focus our question on a specific type of knowledge. (See Appendix IV for a list of potentially useful taxonomies for SoTL studies.) For example, Bruff (2015) found that pre-class reading quizzes consisting entirely of *computational* questions prepared students better for class than quizzes consisting entirely of *conceptual* questions. He noted that his study had treated all his students as a single type, so it was possible that conceptual questions might be more effective for some subgroups of students. Another way to narrow a question on pre-class reading would be to focus on student approaches to understanding just the definitions they encounter in the reading or how they interact with the examples they read.

Considering how the knowledge of the subject matter is organized (whether by experts in the discipline or by novices) is yet another potential approach to refining questions. For example, in refining the question, "How do students read the textbook (or journal articles) for this class?" we might want to consider how students organize information gained from reading. Perhaps the root of our teaching problem might be that students are unprepared to organize new information effectively. Similarly, for the question, "Why are students unable to begin effectively when encountering multi-step problems?" we might refine the question to ask what organizational structures students bring to multi-step problems or how such problems have been (or should be) scaffolded by the instructor.

> **Pause: Narrowing—Questions to Consider**
>
> - What terms or phrases in my question need clarification? How do I want to define these terms more precisely?
> - Is the disciplinary concept or topic composed of smaller ideas or subtopics? Would restricting my investigation to just one of these be an appropriate narrowing?
> - Is there a taxonomy or classification of knowledge or learning (general or disciplinary) that I might use to refine my question?
> - Should my question pay particular attention to certain subgroups of students?
> - Do I want to investigate my question in a single section of a course, in multiple sections, or in a broader set of courses?

Consulting the literature can also help narrow a question. We take up this topic as part of **grounding a question**.

Grounding a Question

After some effort to clarify and narrow our question, we should ground the question in the literature, institution, and discipline. Narrowing the question first can make the grounding process more efficient. And then again, the grounding process may suggest a further narrowing or lead us to shift the locus or focus of the investigation.

Doing a Literature Search

When getting started on a SoTL investigation, a common pitfall is to postpone or fail to check what published work in SoTL or DBER says is already known about the question. There are many reasons why it is valuable to seek out the relevant literature early in the project. Knowledge of what others have done can help formulate the project question, improve the design of the project, provide ideas or instruments for measuring outcomes, and save repeating the work of others. Axtell and Turner (2015) attested to this: "We initially viewed the literature review as an onerous task to be completed for form's sake, but we soon realized that the literature review was playing a central role in shaping our study question and our thoughts for how to design the study" (139). Wagstrom (2015) described how her literature search prompted her to change the site of the investigation from a college algebra course to the prerequisite for that course.

Some suggestions for doing a literature search follow. Begin by expressing the topic of interest in two or three different ways using synonyms. Those words or phrases become the search terms. To get an idea of how effective these search terms are try them out with Internet search engines such as Google or Google Scholar. If the result is too many or too few hits, refine the inputs. For too few hits, use a wildcard, usually an asterisk (*), at the end of words that can have multiple endings to search for all the words simultaneously. For example, using "learn*" will capture the terms "learn," "learners," and "learning." The logical connector "OR" can produce more results since the search "gender" OR "equity" will find documents containing either word. For too many hits, do "AND" and "NOT" searches, or limit the search to certain fields, such as the abstract or title. Next turn to the Education Resources Information Center or ERIC (https://eric.ed.gov), the world's largest digital library of education literature, and PsychINFO (http://www.apa.org/pubs/databases/psycinfo/index.aspx), a database of the American Psychology Association. In these databases enter the search terms into the "KeyWord," "Descriptor," and "All Text" fields as the results may vary.

Although ERIC dominated the educational research scene up until the early 2000s, now there are more database options available to researchers from various publishers (EBSCO, Gale, etc.). The campus librarians can help investigators to determine if their campus subscribes to these or other databases, such as Education Full Text. These databases and Google Scholar are more likely than ERIC to cross disciplinary lines in the results they generate.

Scan the abstracts of the results from the searches to see if they seem relevant. For the ones that do, take note of the descriptors, keywords, and authors' names. Try new searches based on these terms. If databases do not provide the full text of the resource, seek out items of interest from other sources such as the campus library, interlibrary loan, or the publisher. Be organized and keep good records of the search, what was found, and why it was of interest. Save the search statements for possible reuse or later modification. Bibliographic management tools such as EndNote or Zotero (for MsWord) or BibTeX (for LaTeX) make it possible to capture the citations in an appropriate format for the bibliography. Be aware that these tools are not 100 percent accurate.

Another tactic to expand the search is to use **citation chaining**, sometimes called a **snowball strategy**, in one of these ways. When reading the sources related to the study, check their bibliographies for older work on the subject. It may also be possible to find newer literature that cites the sources already identified by using resources such as *Web of Science* (http://webofknowledge.com). Check with campus librarians to see if there is access to this or a similar resource. Finally, from the literature, identify scholars who have done recent work on the topic and contact them to describe the study and inquire if they know of newer unpublished work.

Searches of disciplinary journals devoted to teaching and learning can be made using usual disciplinary search methods to find appropriate articles. However, articles of interest may also appear in general SoTL journals and higher education publications. The Center for Excellence in Teaching and Learning at Kennesaw State University maintains a list of disciplinary specific journals (including various STEM fields) and general or interdisciplinary journals on teaching (https://cetl.kennesaw.edu/teaching-journals-directory), most with a link to the journal's website.

When doing SoTL, it may be helpful to go beyond articles in disciplinary, education, or SoTL journals and look at edited volumes of SoTL work from multiple disciplines. Examples of these are: Chick et al. (2012); Ferrett et al. (2013); Gurung et al. (2009), Huber and Morreale (2005); and McKinney (2012). Then there are books on topics that transcend disciplines such as critical reading (Manarin et al. 2016), identifying bottlenecks to student learning in a discipline and strategies that will help students work through these obstacles

(Pace and Middendorf 2004), or threshold concepts that fundamentally transform student learning and understanding of a discipline (Meyer and Land 2003). Talking with colleagues in other disciplines is another way to learn about resources and relevant literature that cross disciplinary boundaries. Discussions with colleagues may lead to collaborations. Projects can benefit from interactions not just with STEM colleagues, but with sociologists, psychologists, and historians too. For more information on conducting a literature search see the tutorial provided by the Taylor Institute for Teaching and Learning at the University of Calgary (Taylor Institute 2016).

Remember, a good literature search informs us about our SoTL question and later on, if we want to publish our findings, it enables us to explain how our question and findings fit with and extend current knowledge. Giving serious attention to the literature review at the outset increases the likelihood of getting the work published (see Chapter 8, Publishing an Article).

Situational Factors

In conjunction with looking at the literature, we should think how situational factors might affect our question. Previous work on the problem may have involved a very different type of institution, class size, educational system, etc. We should consider whether the question, or information derived from the work, should or will change based on these factors. For example, a flipped classroom in a 300-student course might operate very differently from a flipped classroom in a 20-student course. We should further consider whether the type of institution or student background will affect the investigation. For example, the literature on first generation college students shows how student background can be relevant in questions of teaching and learning.

Situational factors may also influence how we can conduct the study or the type of evidence we can collect. For example, a campus that requires students to have a specific electronic device creates the possibility of designing a study that gathers data using the device. The military academies provide a particularly striking example of a special situational factor. Because they can assign students to different sections of the same course, studies can be conducted with random assignment of subjects to experimental and control groups. This is rarely possible at most other institutions.

We should think about our intended subjects. Some questions might focus on students with certain characteristics, but we can't always guarantee that the class enrollment will provide the types of students we want to study.

For example, a question exploring the differences between transfer students and continuing students might encounter this situation.

How students take advantage of pedagogical changes is often not in the way we expect. Kirschner's (2002) work on **student affordances**, that is, how students make use of innovations versus how we intend them to be used, suggests another situational factor. For example, a project on estimation was derailed because the class had a large number of engineering students with access to AutoCAD, leading them to believe they could answer the question exactly.

The culture of an institution can affect projects and questions in other ways too. Community-based learning projects may function very differently and require different study methods at a small faith-based college versus a large public university. Recognizing these potential issues at the question-framing stage can allow us to refine question in ways that take advantage of the local situation. Sometimes considering situational factors will help us recognize that a related *What is?* question is more appropriate and more interesting than the *What works?* question we were framing.

Disciplinary Factors

The discipline of a study may influence the question. While the chasm that exists between experts and novices transcends disciplines, expert (or novice) behavior is often discipline specific. For example, the approaches used to understand a text may vary from discipline to discipline. Consequently, a researcher looking at student reading of texts in one area might do well to tease out these differences if their students come from a variety of majors.

The method used to generate new knowledge in a field also differs by discipline. For example, mathematicians use proofs as a method for understanding conjectures and refining them into theorems. This often involves discovering appropriate definitions to produce a valid argument. Thinking about how disciplinary researchers try to answer questions and understand concepts in the discipline can help to focus a problem on those factors. For those considering disciplinary factors as a way to refine a question, we recommend the methods for decoding a discipline (Pace and Middendorf 2004; Pace 2017) and the work of Meyer and Land (2003) on threshold concepts that create bottlenecks in student learning. We discuss these two frameworks in more detail in Chapter 3 (Seeking to Understand or Explain). While SoTL methodologies do not have to mirror those of the disciplines, disciplinary methods can help frame a question, and further they can suggest what aspects of the question could be investigated.

> **Pause: Grounding—Questions to Consider**
>
> - What does the disciplinary and educational literature say about my question?
> - Does my literature search suggest I should change my question to study another aspect of the problem?
> - Does class size or type of institution matter?
> - Who will the subjects be and do their backgrounds matter?
> - Should my question focus on subgroups of students with particular characteristics?
> - Should my question be reframed as a *What is?* question because of situational factors?
> - Will I be able to measure what I want in my situation?
> - Are there student affordances and, if so, do they matter?
> - Would attending to unique characteristics of the discipline inform my question?
> - Do the disciplinary practices of experts and novices play a role in my question?
> - What method of study will be acceptable to my disciplinary colleagues?

Identifying Assumptions

Another step in refining a question is to identify the assumptions that underlie the question itself. For example, when instructors begin to ask questions about students reading texts ineffectively, the first assumption some of them make is that the students don't do the reading, and other assumptions might be made about why they don't do so. When Uri Treisman (1992) was studying what factors led to African American students underperforming in calculus at the University of California, Berkeley, his colleagues started with many assumptions about why those students underperformed, all of which turned out to be incorrect. By questioning and investigating those assumptions, Treisman found that it was ineffective learning strategies, social isolation, and other nonacademic factors that had a significant negative impact on the students' learning, not poor study habits or lack of background knowledge. Once he had identified these factors, he was able to develop an alternate learning environment in which minority students excelled and thrived.

The assumptions underlying a question are important to recognize as they often implicitly drive the question in one direction. For example, the *What works?* question: "Do students attain the same level of content mastery in an active learning class?" suggests that the researcher may have these underlying assumptions: the current level of content mastery is important, the current content mastery is the most appropriate content mastery, and active learning improves other types of learning. Assuming these are the researcher's assumptions, they give rise to several possible refinements of the question. It may be the case that by explicitly identifying an assumption regarding current levels of content mastery that the researcher decides that this is not what is truly of interest. Moreover, recognizing the second assumption may lead to refining the question by first describing what content mastery students achieve, thus opening up the SoTL project to investigate what else might be achieved through active learning.

After identifying our assumptions, we should check whether they are true. We can see whether there is research supporting them, and if not, we should contemplate what would happen in the study if the assumptions were false. Careful consideration may lead to a reformulation of the project or even to a different project, one that tries to determine whether or not the assumptions are correct. Questioning assumptions we might have made about what students do outside class can be particularly important. We may be tempted to think that we know how students act because we were once students, but upon reflection we will recognize we are not our students.

> **Pause: Assumptions—Questions to Consider**
> - What assumptions am I making about the students?
> - What assumptions am I making about the topic of the study?
> - What effect will it have on the project or question if my assumptions are incorrect?
> - Is it possible to test these assumptions at the start of the project, and, if so, should I?

Feasibility

After the grounding our question in the literature, looking at disciplinary and situational factors, and examining assumptions, we need to think about the feasibility of investigating our question. Suppose we want to know how a course for future secondary science teachers that involves students writing lesson plans with active learning techniques will influence their future teaching. Unless we have a way of getting information about their instructional practices once they are teachers, our question will need to be revised. So, before finalizing our research question, we should consider the extent to which our available time, expertise, and resources (including student participation) are sufficient to carry out the project.

> **Pause: Feasibility—Questions to Consider**
> - Can reasonable progress be made on the question in the time available?
> - Are there sufficient resources available?
> - Will students be willing participants?
> - Is it possible to measure what is desired?

A negative answer to any of the feasibility questions suggests that it may not be realistic to undertake the study without first revising the research question. Other feasibility issues may arise later as the project progresses to the research design (see Chapter 3) or data collection and analysis (see Chapters 4–7). No matter how interesting an answer to our question would be, if there is no reasonable way to collect meaningful data, or there is no way to evaluate the data we collect, the question will need to be restructured. If we do not have the necessary expertise for all aspects of the study, we can seek collaborators (see Chapter 8, Finding Collaborators and Support). Sometimes feasibility can be tested with a small pilot study.

Revisiting Goals

After having gone through all the steps to refine our question and check for feasibility, it is time to look back and make sure the refined question is the one we want to investigate or if there might be a related question we are more interested in. If we started with a *What works?* question, we might now realize that information from a *What is?* question would help us discover what works. Thus, we should check whether our goals have changed and make sure our refined question is still interesting to us (and to others). Sometimes we will decide to go through the refining process again. To recap, the ways to refine a question are:

1. Identify our teaching problem.
2. Generate an initial question.

3   Narrow the question.
4   Discover what is known.
5   Examine the situational factors.
6   Examine the disciplinary factors.
7   Examine assumptions.
8   Check feasibility.
9   Revisit our goals to see if we want to revise the question.

SoTL scholars refine their questions all the time, sometimes even during the course of the study, because of the way things developed. While this might not be ideal for reproducibility, little about studying teaching and learning is.

REFERENCES

Aurigemma, J., S. Chandrasekharan, N.J. Nersessian, and W. Newstetter. 2013. "Turning Experiments into Objects: The Cognitive Processes Involved in the Design of a Lab-on-a-Chip Device." *Journal of Engineering Education 102* (1): pp. 117–40. doi: 10.1002/jee.20003.

Axtell, M., and W. Turner. 2015. "An Investigation into the Effectiveness of Pre-Class Reading Questions." In *Doing the Scholarship of Teaching and Learning in Mathematics,* edited by J. Dewar and C. Bennett, pp. 137–44. Washington: Mathematical Association of America.

Bass, R. 1999. "The Scholarship of Teaching and Learning: What's the Problem?" *Inventio, 1* (1): pp. 1–10.

Bruff, D. 2015. "Conceptual or Computational? Making Sense of Reading Questions in an Inverted Statistics Course." In *Doing the Scholarship of Teaching and Learning in Mathematics*, edited by J. Dewar and C. Bennett, pp. 127–36. Washington: Mathematical Association of America.

Chick, N., A. Haynie, and R. Gurung. 2012. *Exploring More Signature Pedagogies: Approaches to Teaching Disciplinary Habits of Mind.* Sterling: Stylus.

Dewar, J., S. Larson, and T. Zachariah. 2011. "Group Projects and Civic Engagement in a Quantitative Literacy Course." *PRIMUS: Problems, Resources, and Issues in Mathematics Undergraduate Studies 21* (7): pp. 606–37.

Elmendorf, H. 2006. "Learning through Teaching: A New Perspective on Entering a Discipline." *Change 38* (6): pp. 37–41.

Felder, R. M., and R. Brent. 1996. "Navigating the Bumpy Road to Student-Centered Instruction." *College Teaching 44* (2): pp. 43–7.

Ferrett, T. A., D. R. Geelan, W. M. Schlegel, and J. L. Stewart, editors. 2013. *Connected Science: Strategies for Integrative Learning in College.* Bloomington: Indiana University Press.

Gurung, R., N. Chick, and A. Haynie, editors. 2009. *Exploring Signature Pedagogies: Approaches to Teaching Disciplinary Habits of Mind.* Sterling: Stylus.

Henderson, C., and A. Rosenthal. 2006. "Reading Questions: Encouraging Students to Read the Text Before Coming to Class." *Journal of College Science Teaching* 35 (7): pp. 46–50.

Holcomb, J. 2015. "Presenting Evidence for the Field That Invented the Randomized Clinical Trial." In *Doing the Scholarship of Teaching and Learning in Mathematics* edited by J. Dewar and C. Bennett, pp. 117–26. Washington: Mathematical Association of America.

Huber, M.T., and S. P. Morreale, editors. 2005. *Disciplinary Styles in the Scholarship of Teaching and Learning: Exploring Common Ground*. Washington: American Association for Higher Education.

Hutchings, P., editor. 2000. *Opening Lines: Approaches to the Scholarship of Teaching and Learning*. Menlo Park: The Carnegie Foundation for the Advancement of Teaching.

Hutchings, P., M. T. Huber, and A. Ciccone. 2011. *The Scholarship of Teaching and Learning Reconsidered: Institutional Integration and Impact*. San Francisco: Jossey-Bass.

Kirschner, P. A. 2002. "Can We Support CSCL? Educational, Social and Technological Affordances for Learning." In *Three Worlds of CSCL: Can We Support CSCL?*, edited by P. A. Kirschner, pp. 7–47. Heerlen: Open University of the Netherlands.

Litzinger, T. A., P. V. Meter, C. M. Firetto, L. J. Passmore, C. B. Masters, S. R. Turns, G.L. Gray, F. Costanzo, and S. E. Zappe. 2010. "A Cognitive Study of Problem Solving in Statics." *Journal of Engineering Education* 99 (4): pp. 337–53. doi: 10.1002/j.2168-9830.2010.tb01067.x.

Manarin, K., M. Carey, M. Rathburn, and G. Ryland, G., editors. 2016. *Critical Reading in Higher Education: Academic Goals and Social Engagement*. Bloomington: Indiana University Press.

McKinney, K., editor. 2012. *The Scholarship of Teaching and Learning In and Across the Disciplines*. Bloomington: Indiana University Press.

Meyer, J.H.F., and R. Land. 2003. "Threshold Concepts and Troublesome Knowledge: Linkages to Ways of Thinking and Practicing." In *Improving Student Learning—Theory and Practice Ten Years On*, edited by C. Rust, pp. 412–24. Oxford: Oxford Centre for Staff and Learning Development.

Offerdahl, E., and L. Montplaisir. 2014. "Student-Generated Reading Questions: Diagnosing Student Thinking with Diverse Formative Assessments." *Biochemistry and Molecular Biology Education* 42 (1): pp. 29–38.

Pace, D., and J. Middendorf, editors. 2004. *Decoding the Disciplines: Helping Students Learn Disciplinary Ways of Thinking*, New Directions for Teaching and Learning: No. 98. San Francisco: Jossey-Bass.

Pace, D. 2017. *The Decoding the Disciplines Paradigm: Seven Steps to Increased Learning*. Bloomington: Indiana University Press.

Rolka, C., and A. Remshagen. 2015. "Showing Up is Half the Battle: Assessing Different Contextualized Learning Tools to Increase the Performance in Introductory Computer Science Courses." *International Journal for the Scholarship of Teaching and Learning* 9 (1). doi: 10.20429/ijsotl.2015.090110.

Taylor Institute. 2016. "Conducting a Lit Review." The Taylor Institute for Teaching and Learning at the University of Calgary. Accessed May 26, 2017. http://sotl.ucalgaryblogs.ca/doing-sotl/conducting-a-lit-review.

Treisman, P. U. 1992. "Studying Students Studying Calculus: A Look at the Lives of Minority Mathematics Students in College." *College Mathematics Journal 23* (5): pp. 362–73.

Wagstrom, R. 2015. "Using SoTL Practices to Drive Curriculum Development." In *Doing the Scholarship of Teaching and Learning in Mathematics,* edited by J. Dewar and C. Bennett, pp. 191–202. Washington: Mathematical Association of America.

Watt, J., C. Feldhaus, B. Sorge, G. Fore, A. Gavrin, and K. Marrs. 2014. "The Effects of Implementing Recitation Activities on Success Rates in a College Calculus Course." *Journal of the Scholarship of Teaching and Learning 14* (4). doi: 10.14434/v14i4.12823.

CHAPTER 3

Designing a Research Study

Introduction

The scholarship of teaching and learning (SoTL) involves the systematic investigation of a question we have about student learning and we look for answers in evidence generated by students. After framing a researchable question, we have to decide how we will go about gathering evidence to answer our question. So, this chapter examines some basic considerations of research design, such as whether or not to have experimental and control groups. In addition, because students are the subjects of investigation, a SoTL researcher needs to obtain human subjects clearance in order to publish the results of the study. We close the chapter with a discussion of this issue.

Challenges of Education Research Design for SoTL

The National Research Council (2002, 51) identified guiding principles for scientific research in education, calling for research that:

- Poses significant questions that can be investigated empirically
- Links empirical research to relevant theory
- Uses research designs and methods that permit direct investigation of the question

The Scholarship of Teaching and Learning, Jacqueline M. Dewar, Curtis D. Bennett, and Matthew A. Fisher.
© Oxford University Press, 2018. Published 2018 by Oxford University Press

- Provides an explicit and coherent chain of reasoning
- Replicates and generalizes across studies
- Discloses results to encourage professional review and critique by peers.

United States (US) government agencies and professional societies have held workshops and produced reports to promote high quality research studies and assist with the evaluation of research proposals in the area of science, engineering, and mathematics education (Suter and Frechtling 2000; American Statistical Association 2007; American Association for the Advancement of Science 2012; Institute of Education Sciences and the National Science Foundation 2013). These reports can provide helpful insights and contain examples of education research projects in science, technology, engineering, and mathematics (STEM) that have been funded by the National Science Foundation. But, designing a study can pose a number of challenges to faculty members interested in doing SoTL, which we now consider.

When attempting to show a causal relationship in response to a *What works?* question, the **gold standard** is the **experimental study** where subjects are randomly assigned to an **experimental group** and a **control group**. Because SoTL research questions often arise from a person's own teaching practice, and investigations are often undertaken in the researcher's own classroom, meeting the gold standard can be difficult or impossible. Moreover, how appropriate random assignment experimental studies are in education research and in SoTL has been a matter of considerable discussion by members of both communities (Berliner 2002; Cook and Payne 2002; Grauerholz and Main 2013; Poole 2013; Shulman 2013). In some ways, the challenge is similar to the one faced by scientists doing ecological research; it is rarely feasible to set up a control group for an entire ecosystem.

Poole (2013) discussed how disciplinary views differ over what is knowable and whether behavioral indicators alone suffice to measure educational outcomes. Grauerholz and Main (2013) described the near impossibility of controlling the factors that can influence the learning environment in a classroom. In a keynote address at an international SoTL conference, Lee Shulman, President Emeritus of the Carnegie Foundation, discussed at some length the idea that in social science and education generalizations decay over time because contexts (for example, the students, the standards, the fields or professions, the way information is accessed, and so on) change (Shulman 2013). Shulman argued that the value of situated studies—the type frequently done in SoTL—should not be diminished, as it is no minor accomplishment to "know a small part of the world as it really is." He reminded the audience that, in 1980, the educational psychologist Lee Cronbach told his colleagues that results of empirical studies in social science can rarely, if ever, "identify 'right' courses of action" (Cronbach 1982, 61). Kagan (2012a, 2012b) has presented similar views about the importance that should be given to context.

Studies that make comparisons between groups that are not randomly assigned are called **quasi-experimental studies.** We begin with one such example undertaken in a multi-section quantitative literacy (QL) course. Five out of 20 sections of the course were revised to include semester-long group projects involving local community issues that students could investigate using the mathematical topics of the course. Students enrolled without knowing that some sections would have projects while others would not, so the assignment was at least blind, but not random. The instructors saw no evidence of students changing sections after seeing the syllabus. The course fulfilled a graduation requirement for all students whose majors did not require a mathematics course. The enrollment in the project-based sections reflected the typical mix of majors and years in school. Student background characteristics (gender, year in school, age, majors, entering SAT scores, and the like) could have been more formally compared, but were not. The project-based experimental group included 110 students and the comparison group had more than twice that. The investigators considered this to be a *What could be?* rather than a *What works?* study. Evaluation of the efficacy of the projects approach included pre- and post-comparisons between the two groups on a test and on an attitudinal survey. In addition, a focus group and a knowledge survey were conducted with the experimental group. We return to this example several times when we discuss the issue of alignment between assessment tools and the outcomes being measured (Chapter 3, Aligning the Assessment Measure with What is Being Studied), knowledge surveys (Chapter 5, Knowledge Surveys), and focus groups (Chapter 6, Focus Groups). Readers who are interested in a full account of the study should consult Dewar et al. (2011).

Dealing with the Lack of a Control Group

In a typical SoTL investigation, finding a comparison group is difficult, if not impossible, for a number of reasons. The most basic is that we may only have the students in our classroom as our subjects and no access to a control group. Yet, as noted in Adhikari and Nolan (2002), investigating a *What works?* question requires some sort of comparison group or, at least, baseline data.

We describe several ways to clear this hurdle, each with its own set of advantages and disadvantages.

It may be possible to teach two sections of the same course in the same semester. If so, one section could be taught using the traditional approach and one using the experimental approach. Of course, as with many forms of classroom research, a question naturally arises concerning the instructor's impartiality. The instructor is trying a new approach that he or she thinks will be better in some way and has formulated a research question about it. Will the instructor's beliefs or interests give the experimental section an advantage? It is natural to consider recruiting an impartial and capable colleague to teach the experimental section. However, if the instructor of the experimental group is not as familiar or comfortable with the teaching method or intervention, the results may not be the same as they would be for an instructor who is. Bowen (1994) and Kaus (2015) both observed that, even if two or more appropriate colleagues can be identified, difficulties can still arise with collecting data, either being able to do so in a reproducible manner or at all. Extra variables such as these that could distort the results are called **confounding factors**. The lack of strict controls when comparing student performance across groups is one factor contributing to the perception that SoTL studies are less generalizable than formal education research studies.

Sometimes, the intervention can be alternated between two sections of a course taught by the same instructor. This method is called a **switching replications design**. For example, the instructor can teach material covered in the first and third quizzes in a course with the experimental approach in one section and with the standard approach in another, and then switch the approaches for the second and fourth quizzes. This can eliminate the confounding factor resulting from different levels of ability in students in the different sections. However, a disadvantage to a switching replications design is that it cannot be used to assess the cumulative effect of a treatment over an entire semester. Also, it might introduce a different confounding factor if one set of quizzes has more difficult material. A switching replications design minimizes potential risk to students' grades and learning, and that may find favor with institutional review boards when they review research proposals to see that they meet ethical guidelines. (For further discussion of these boards, see the Human Subjects Considerations section in this chapter.) However, when choosing to alternate an intervention between two sections, instructors may need to consider whether students could possibly attend the other section or study with students who do. This method can be more vulnerable to such crossover students, since it measures shorter-term effects rather than those developed over an entire course.

In a paper presented at the 2012 Canadian Association of Physicists Congress, Morris and Scott (described in Healey et al. 2013) wanted to know if access to a formula sheet on physics examinations would enhance, detract, or make no difference in the resulting scores. They tried to design a study that would produce evidence that would be scientifically rigorous and credible to their colleagues while minimizing the potential risks to their students. They were able to achieve this through a careful switching replications design, and close cooperation between the two teacher-researchers. To obtain enough student participants to have the potential to achieve statistically significant results, they studied two concurrent sections of the same course.

All students, whether participating in the study or not, wrote the first midterm with a formula sheet, as per usual. Participating students wrote one remaining midterm with a formula sheet and one without, depending upon their group assignment. Non-participants wrote the second midterm with a formula sheet, but were given the option of writing the third midterm with or without a formula sheet. For all students, the lowest midterm mark was dropped in the calculation of final grades. In this way, the instructors ensured that whether using a formula sheet during the midterm exams was beneficial or not, students would be exposed to no additional risk compared to the traditional organization of the class, "thus meeting the *pragmatic* goal of managing potential risk to participating and non-participating students" (Healey et al. 2013, 30).

While they found no significant difference in overall course performance, they were led to a hypothesis for a new SoTL study: "Do students score higher on conceptual questions when they do not use a formula sheet?"

Another approach to the lack of a control group when trying a new method is to rely on comparison data obtained from **prior cohorts** (Hutchings 2002). Suppose a faculty member thought a general education physics course where students designed their own labs would give students a better understanding of what it means to "do physics" than the general education physics course with traditional labs. The instructor could teach the traditional general education course first and gather baseline data through tests, surveys, and perhaps reflective writing. This would provide data from a "control" group for comparison with the data later gathered using the same instruments when teaching the revised course.

The next example describes a different situation, one involving the development of a totally new course to try to improve retention in the major. In this case, the course developers found a way to make some comparisons to prior cohorts. In 1992, the Mathematics Department at Loyola Marymount University introduced a workshop course for beginning mathematics majors. It aimed to enhance their education in a number of ways, including developing skills in problem solving and mathematical writing, thereby increasing the likelihood of success in the major. To increase student motivation to persist in the major, the course also discussed mathematical people and careers to show the human side of the discipline and inform students of the variety of rewarding career paths open to mathematics majors. Several assessments of the course's effectiveness (Dewar 2006) utilized comparisons with prior cohorts of mathematics majors. At the end of the 1992 introductory year, a survey about mathematical careers and mathematicians showed the first group of freshmen taking the course to be more knowledgeable than more advanced students who had not taken the course. Over the next several years, data gathered allowed comparison of the retention rates of the first few cohorts with rates prior to the introduction of the course, showing the dropout rate was nearly halved. Other assessments involved pre- and post-tests of problem solving, a portfolio assignment that demonstrated improvement in writing from the beginning to the end of the course, and self-reported improvements in confidence at the end of the course, none of which made comparisons to prior cohorts. Overall, the results were so convincing that the course became a requirement in the first year of the major.

Comparisons between an experimental and a control group can raise ethical questions. We hear of medical trials being interrupted because the results for the treatment group are so positive that it is deemed unethical to withhold the treatment from the placebo group. A similar sort of question, though not of the same life-threatening nature, can arise in SoTL studies. Is it fair to provide only some students with special learning experiences, especially if the instructor truly believes they make a difference? Faculty, particularly those at small mission-driven liberal arts colleges committed to providing students with personalized, high-quality learning experiences, may consider it unethical to perform treatment and control group SoTL studies (Gieger et al. 2015).

A different sort of ethical question can also arise in doing SoTL work. Is it ethical to ask students to perform tasks they were not prepared for? For example, is it fair to ask students taught in a traditional elementary statistics course to analyze real data and provide a written report if these tasks were not part of their course, but were the focus of the instruction in the section being studied in a SoTL investigation? John Holcomb (2002) faced this dilemma and discussed how it influenced the direction of his investigation as a Carnegie scholar.

Although, as a statistician, he found it difficult to "let go of the control group model" (21), he did not use one. During the semester, his students worked collaboratively on homework assignments involving real data sets. His outcome measure was their performance on similar problems on take-home midterm and final exams that were completed independently.

Another practical and ethical issue that can arise in SoTL studies is how much extra work participating in the study or its assessments will require of the students. The SoTL researcher should aim to place little or no extra burden on students. By giving careful attention to design, many assessment tasks intended to capture SoTL evidence can be embedded within course assignments. Ideally, much of the evidence gathered will emerge organically from learning activities or evaluation processes built into the course. See Chapter 4 (Assignment Design) for suggestions on how to do this.

An alternate way of showing the efficacy of a teaching method or intervention is to measure achievement of stated learning goals. This is known as a **criterion-referenced evaluation** and it avoids the ethical dilemma inherent in making comparisons by assessing two groups of students using tasks for which only one group received instruction. A criterion-referenced approach requires specifying in advance the learning goals and designing assessments that measure them. Typically, this involves pre- and post-comparisons of the students taking the course. This approach comes with its own set of concerns, including the importance of aligning the assessment instrument with the outcomes that are to be measured (see Aligning the Assessment Measure with What is Being Studied, in this chapter).

To Seek Statistical Significance or Not

SoTL researchers often wrestle with the issue of whether or not to seek to show statistically significant differences because their investigations take place in classrooms where the number of students involved is too small to seek **statistical significance** using common comparison methods such as *t*-tests. Some researchers will teach the experimental version several semesters in a row to aggregate enough data to meet sample size requirements. Others reject this approach because they know they will find it impossible to resist making adjustments to the course during the several semesters it would take to gather sufficient data.

Sometimes, unanticipated attrition in a course can sabotage plans to use pre- and post-course comparisons to show an effect. Collecting pre- and post-data from one component of a course rather than over an entire course may provide sufficient data for statistical comparisons. Time spent early on considering all the things that could go wrong with a project is time well spent. Beginning with a small-scale version of the study, a **pilot study**, is one way to detect problems with the design or with the instruments or measures that will be used to obtain data.

STEM instructors who want to embark on a SoTL project but lack a background in statistics need not be too concerned. Many worthwhile projects, especially those belonging to the *What is?* category of the SoTL taxonomy, require no inferential statistics. We give some specific examples in Chapter 7 (Introduction).

Researchers new to SoTL should be aware that, although ***p*-values** have been central to statistical data analysis for decades and lie at the heart of a claim of "statistical significance," they have come under scrutiny from the social sciences. *Basic and Applied Social Psychology* became the first social science journal to ban articles that relied on null hypothesis significance testing procedures (Trafimow and Marks 2015). Subsequently, the American Statistical Association (Wasserstein and Lazar 2016) issued a statement on the limitations of *p*-values, noting: "While the *p*-value can be a useful statistical measure, it is commonly misused and misinterpreted" (131). The statement contains six principles regarding the use of *p*-values as guidance for researchers who are not primarily statisticians. The following caveat is from the fifth of those:

> Statistical significance is not equivalent to scientific, human, or economic significance. Smaller p-values do not necessarily imply the presence of larger or more important effects, and larger p-values do not imply a lack of importance or even lack of effect. Any effect, no matter how tiny, can produce a small p-value if the sample size or measurement precision is high enough, and large effects may produce unimpressive p-values if the sample size is small or measurements are imprecise (132).

Later in this chapter (Which Measure of Change?) we discuss other ways of measuring the impact of an intervention, including normalized gain and Cohen's *d* effect size, which can be used instead of or in addition to a statistical test for significance.

Aligning the Assessment Measure with What is Being Studied

To be useful in a SoTL study, an assessment measure must align with, that is, be able to measure, some aspect of what is being studied. The following cautionary tale shows that this might not always be as obvious as it seems. Recall the example of the QL course that Dewar et al. (2011) revised to include a civic engagement component where students applied the elementary mathematics they were learning to semester-long projects involving campus or community issues. On nine multiple-choice questions, when the pre- and post-results in the projects-based sections were compared to those in sections of the course without the projects, there was a difference on only one question, namely, one that involved interpreting the meaning of a "margin of error." Achieving a better outcome on only one out of nine questions was a disappointing result that prompted serious reflection. One possible explanation was that the standard approach to teaching QL was also doing a good job of teaching content. The projects approach was not better at improving skills but was better at improving attitudes and understandings of how mathematics connects to and is useful in daily life.

Later, the well-known physics education researcher (and Nobel Prize winner), Carl Wieman, paid a visit to campus. In a discussion about this SoTL project he pointed out that the other eight pre- and post-test questions might not be designed to find any distinctions. He was right. All the other questions were straightforward computational problems such as "What is the monthly payment on a car loan if…?" They tested factual knowledge and the ability to apply formulas and perform computational tasks. The addition of research projects on community issues could not be expected to have an effect on those skills.

The question that showed a significant difference in performance favoring the projects-based approach was not testing facts or procedures. It required an evaluation of which of four given explanations for the meaning of the phrase "margin of error" was the best. Evaluation is a higher order task in Bloom's taxonomy (Anderson and Krathwohl 2001). All QL students, not just those in the projects-based sections, computed margins of error on homework and studied the procedure in preparation for quizzes and tests. Of course, the meaning of the margin of error appeared in the text and should have been discussed in all classes. However, in the projects-based sections many of the groups chose projects that involved gathering survey data. They had to compute the margin of error for their data and they had to explain what it meant in their own words in their written reports and in-class presentations. Of all the pre- and post-test questions, only the conceptual question on interpreting the margin of error could possibly capture any improvement between the two approaches to teaching QL. This was so obvious once Carl Wieman asked whether the other pre- and post-questions had the potential to detect a difference.

Which Measure of Change?

When using pre- and post-comparisons, some consideration needs to be given to which measure of change to use. Possibilities include actual change, percent change, or **normalized gain**. The last of these was popularized by Hake (1998) in his large-scale study of interactive-engagement (active learning) teaching methods in introductory physics. Normalized gain, the ratio of actual gain to maximum potential gain, is defined as,

$$g = \frac{post - pre}{100 - pre},$$

where the *pre*-test and *post*-test each have a maximum score of 100.

This measure of change was developed to allow comparisons of student populations with different backgrounds. Students who begin with significant knowledge about a topic, and therefore score well on a pre-test, are unable to show much actual or percent change on a post-test. This measurement difficulty is resolved by using the normalized gain, since it represents the fraction of the available improvement that is realized. To avoid inflating the gain by including the pre-scores of students who stopped attending or dropped the course, normalized gain should be calculated with **matched data**, that is, data for students with both pre- and post-scores. In physics education research, the labels (and corresponding ranges)—small (less than 0.3), medium (0.3 to 0.6), and large (greater than 0.7)—are generally accepted and treated as meaningfully different (McKagan et al. 2017). The normalized gain has become so widely used in physics education that its properties have been extensively studied. For example, Stewart and Stewart (2010) have shown that it is insensitive to guessing.

In education research outside of physics, it is more common to report an **effect size** than a gain. Effect size is also a measure of the difference between two groups. In a *What works?* SoTL study, the **absolute effect size** would simply be the difference of means between the treatment and control group or between the average pre- and post-scores. This indicates *how much* the treatment or instruction affected the students. Depending on the variable under consideration, it may or may not be clear whether an absolute effect size is a lot or a little. If an intervention raises the average of the final exam scores by 30 percentage points, that is clearly a lot! On the other hand, when considering the averages of a five-point Likert scale response from 1 (corresponding to a response of "Strongly Disagree") to 5 ("Strongly Agree") on a survey question, is a 0.3 difference meaningful or inconsequential?

A widely-used measure, called **Cohen's d**, describes the effect size by comparing the difference of the means to the standard deviation of the pooled data sets. The formula is simple to compute:

$$d = \frac{m_1 - m_2}{SD_{pooled}},$$

where m_1 is the mean of the treatment group, m_2 is the mean of the control group, and SD_{pooled} is the standard deviation of the pooled data for the two groups. Cohen (1988) provided 0.2, 0.5, and 0.8 as guidelines for interpreting the resulting value as small, medium, and large, but cautioned that these labels are relative not only to each other but "to the specific content and method being employed in any given investigation" (25). One difference between normalized gain and effect size is that effect size takes into account the size of the class and the variance in the scores, but normalized gain does not. According to McKagan et al. (2016),

> Normalized gain helps account for the effect of differing pre-test levels, which allows us to compare courses with very different pre-test scores. Effect size helps account for the effect of differing sizes of error, which allows us to compare courses with different levels of diversity in scores and class sizes. It is statistically more robust to do the latter.

Instructors who lack a background in statistics should not be discouraged by this discussion. As we explain in the next section, there are ways to approach SoTL investigations that require little or no statistics. Another option is to seek out a collaborator in STEM or in the social sciences who is familiar with statistical techniques. In Chapter 8 (Finding Collaborators and Support) we describe how valuable working with colleagues can be.

Seeking to Understand or Explain

A SoTL inquiry can be approached from the perspective of anthropology rather than undertaken as an experimental study. This may involve seeking to understand the causes of the problem, attempting to describe accurately the phenomenon, student behavior, or (mis)understanding that drew attention to the question, or simply documenting student perceptions. In other words, the researcher engages with a "teaching problem" by investigating *What is?* not *What works?* questions. This type of question seeks to make "invisible learning" visible (Bass and Eynon 2009), an endeavor the three authors of this book agree is extremely valuable. In fact, issues that we have been interested in investigating as SoTL scholars have involved *What is?* questions at least as often as *What works?* questions. If it is not feasible to seek statistical significance for a *What works?* question, it might be possible to transform it into a related *What is?* question.

What is? investigations may require the researcher to examine closely what a student does or understands and may involve careful analysis and categorizing of written responses to open-ended questions or assignments or of transcripts of student discussions during their work on tasks. Or they may rest on gathering student perceptions and opinions through surveys or focus groups.

Two relatively new approaches that provide very powerful frameworks for systematically exploring *What is?* questions are **decoding the disciplines** and **threshold concepts**. Decoding the disciplines (http://decodingthedisciplines.org), a process originally developed by David Pace, Joan Mittendorf, Arlene Diaz, and Leah Shopko (Pace and Middendorf 2004; Pace 2017), focuses on narrowing the gap between expert and novice by identifying **bottlenecks** to learning and what experts do to avoid getting stuck at these bottlenecks. For example, Zolan et al. (2004) identified the following as bottlenecks in genetics and molecular biology: students have difficulty visualizing chromosomes, appreciating the distinction between similar and identical chromosomes (i.e., homologs and sister chromatids), and predicting their segregation patterns during mitosis and meiosis.

Meyer and Land (2003) provided another basis for investigating student understanding: **threshold concepts**. These are core concepts in a subject that, when understood, transform an individual's understanding of that subject. Examples include the limit concept in mathematics (Meyer and Land 2003), the ability to think correctly about scale in biology (Ross et al. 2010), and atomicity in chemistry (Talanquer 2015).

A *What is?* approach can produce results that are useful in later attempts to formulate and assess a method to address the problem. A detailed description of how this happened in a study of the use of reading questions in a statistics course for engineers can be found in Bruff (2015). Results of *What is?* investigations can also point to new questions worthy of investigation.

On the other hand, whenever a researcher is successful in showing a statistically significant result with a certain approach, the question of why that approach is better will also be of interest. Answering *why* something "works" in one situation but not in another often involves gathering a different kind of evidence,

evidence that is more descriptive in nature so as to explore "*What is?*" under one set of conditions but not under another.

As discussed in Chapter 2 (A Taxonomy of SoTL Questions), a *What could be?* investigation seeks to describe and understand what happened in a particular situation. This may involve the use of statistics to present evidence of significant change in some aspect of learning along with explanations or reasons why this might have happened. The explanations often emerge from data gathered through interviews, surveys, focus groups, reflective writing, or other methods designed to probe reasoning, understanding, or motivation. Many of the methods for gathering and analyzing evidence for *What is?* and *What could be?* investigations may be unfamiliar to faculty trained in the STEM disciplines. Chapters 6 and 7 focus on these methods.

> **Pause: Research Design—Questions to Consider**
>
> - Which type of question do I have—*What works? What is? What could be?*
> - Am I able to do an experimental study with a random assignment to treatment and control groups?
> - Is there a prior cohort that I could use to make comparisons?
> - Will I make pre- and post-comparisons over an entire course or a portion of a course?
> - Will I have a sufficient number of subjects to seek statistical significance?
> - What variables will influence the study and which ones can I control?
> - Do I want to understand or explain instead of (or in addition to) comparing?
> - Should I reframe my question as a *What is?* question?
> - Do I need to learn methods for gathering and analyzing data?
> - Does the data I plan to collect align appropriately with my research question?
> - If I am trying to measure change, have I chosen an appropriate way?
> - Should I seek a collaborator who can help with statistical or other data collection and analysis methods?

Human Subjects Considerations

At the outset of a SoTL investigation, if the goal is to publish the results, then **human subjects** issues will arise. These may be unfamiliar to many STEM faculty members although life science faculty may have encountered related issues with animal use or human health research.

According to the United States Federal Code of Federal Regulations, a human subject is a person about whom an investigator (whether faculty or student) conducting research obtains data through intervention or interaction with the individual or identifiable private information (32 CFR 219.102.f). Because of past abuses of human subjects in medical trials in populations such as prisoners or minorities in the armed forces, the US federal government has developed procedures requiring **informed consent** for human subjects research. Obtaining informed consent from the subjects of a SoTL study may involve getting written permission from students to collect and use their data in the study. In addition, special rules for obtaining informed consent apply to any subject under the age of 18, a situation that can be encountered in SoTL studies of first-year college courses.

Many other countries and international organizations also have standards and regulations that govern research on human subjects. The Office for Human Research Protections in the U.S. Department of Health and Human Services compiles a list of standards for human research around the world. The 2017 edition contains over 1,000 laws, regulations, and guidelines that govern human subjects research in 126 countries, as well as standards from a number of international and regional organizations (U.S. Department of Health and Human Services 2017). For example, in Canada, in 2001, three federal research agencies jointly created the Interagency Advisory Panel on Research Ethics as part of a collaborative effort to promote the ethical conduct of research involving human participants. The Panel on Research Ethics develops, interprets, and implements the *Tri-Council Policy Statement: Ethical Conduct for Research Involving Humans* (Panel on Research Ethics 2014). All institutions of higher education in Canada that receive funding from the Canadian government are required to follow this policy statement on human research and have local Human Research Ethics Boards (HREB) review research proposals.

Power relationships are inherently part of collegiate-level instruction and they come to the fore in SoTL because the human subjects are frequently the researcher's own students. The Taylor Institute for Teaching and Learning at the University of Calgary, Canada, has published a guide on ethical issues in the scholarship of teaching and learning (Fedoruk 2017). It interprets relevant principles of the *Tri-Council Policy Statement* for researchers conducting SoTL studies and draws on the scholarly literature on research ethics and SoTL. A shorter overview of the ethical considerations involved in doing research on one's own students is contained in an informational document prepared by the HREB at Mount Royal University (Mount Royal University Human Research

Ethics Boards 2012). It advises instructors to take care that students do not feel unduly influenced to participate in the project, to think carefully about whether any conflict of interests might arise from taking on the dual roles of teacher and researcher, and to reveal potential conflicts in their research proposals. While these two documents are intended for SoTL researchers in Canada, they consider overarching principles that should guide all SoTL research. For the remainder of this section, we shift our focus to procedures and considerations for human subjects research in the US.

Because SoTL publications may involve making our students' work or views public, we must follow institutional guidelines for working with human subjects. Most US colleges and universities have a committee or group, typically called an Institutional Review Board (IRB) or a Human Subjects Review Board, charged with ensuring that the US Policy for the Protection of Human Subjects (45 CFR46) is observed. Researchers planning a study involving human subjects must submit an application for review and approval by the local IRB.

Human subjects researchers are expected to inform their subjects of the risks of their involvement in the study and obtain written consent for their participation. On some campuses, studies that involve little or no physical or emotional risk to the subject, and will not reveal anything about the subject's behavior that would be damaging if it became public, may be exempted from the requirements of the Policy for the Protection of Human Subjects. Whether the researcher thinks the study qualifies as **exempt** or not, only the campus review board is authorized to make that decision. Consequently, the researcher must contact the review board and follow its guidelines and procedures.

However, best practice from an ethical perspective may go beyond what an IRB might mandate. Since the work students do for a course is their intellectual property, we really should ask for their consent to use the work in our study, just as we would ask a colleague for consent to use something that was unpublished.

Studies that present no more than a minimal risk to participants may qualify for an **expedited review,** in which the IRB chairperson alone or one or more experienced reviewers designated by the chairperson perform the review. Thus, the study can be approved without waiting for the entire board to convene. To qualify for an expedited review, the study must meet the requirements set forth in the Code of Federal Regulations (45 CFR 46.110). The campus IRB is the best source of information about whether these regulations apply to the proposed study. We strongly caution that some boards use the term "exempt" to describe an expedited review, so be careful about the meaning of the term as used by the local board.

As part of the IRB process, the researcher will typically be asked to describe the study's goals, methods, the intended subjects, and what safeguards will be put into place to ensure their safety and anonymity or confidentiality, and the confidentiality of any data collected. Other questions likely to be on the IRB application include: how subjects will be recruited; what risks or benefits subjects will encounter; and if students are to be involved in the study as co-investigators, how they will be trained and supervised. In addition, IRB applications usually request copies of the form that will be used to obtain a student's written informed consent to participate in the study and of any surveys that would be given as part of the study. We know from experience that very similar institutions can have very different IRB application forms. This underscores the need to consult with the local campus board to find out its particular procedures and requirements.

Typically, the researcher must demonstrate familiarity with the federal requirements for the protection of human subjects, usually by completing an online tutorial program. Certification programs in research ethics include *Protecting Human Research Participants*, available free through the National Institutes of Health (https://phrp.nihtraining.com), and *Protection of Human Subjects* (http://www.citiprogram.org), offered as a subscription service by the Collaborative Institutional Training Initiative. The time required to complete the tutorials varies from one to several hours. Once obtained, the certifications are usually good for a number of years. How long a certification is considered valid and which tutorial programs are recognized will vary from campus to campus. Proof of each investigator's certification is usually required as part of the IRB application. If students will be involved as co-investigators in the study, then they too will have to be certified, and the faculty member needs to make sure they understand and follow the various protocols for protecting student privacy.

In theory, the IRB process must be completed before collecting any data, but if there are existing data, they may be usable. Review boards may grant permission to work with data collected before applying for permission to work with human subjects if the data were gathered as part of normal educational practices, but they may also refuse to allow use of that data.

Because specific details and policies vary, it is essential to become familiar with the procedures and requirements of the local review board. Colleagues on campus in education or psychology are likely to have experience with the process. Asking them for advice is one way to start up a collaboration, which can have numerous advantages when doing SoTL (as discussed in Chapter 8, Finding Collaborators and Support).

> **Pause: Human Subjects—Questions to Consider**
> - What is the IRB application process?
> - When are applications accepted?
> - What is a typical turn-around time?
> - Do I need to find a colleague who is familiar with the IRB process at my institution?
> - Have I completed the required human subjects certification?
> - Do I have a timeline for completing my IRB application far enough in advance so that I will have approval in time to begin collecting my data as planned?
> - If I have students as co-investigators, have I seen to it that they are certified and are aware of the need to protect the privacy of the student subjects?

In closing this chapter, we note that ethical issues have the potential to arise in SoTL studies in several ways, ranging from the research design to the recruitment of subjects, how well informed they are, how freely they can give their consent, and how careful we are to maintain their confidentiality. We encourage those engaging in SoTL to view research ethics as more than getting human subjects approval. We join McLean and Poole (2010, 8) in urging potential scholars of teaching and learning to consider the following:

1. Present the potential costs and benefits to students frankly, making explicit reference to one's position of authority where appropriate, and acknowledging the degree of uncertainty regarding the full range of impact on students' educational experiences.
2. Ensure that the "social penalties" arising from the choice to participate or not participate are minimized, if not eliminated, by reducing the public nature of the decision not to participate and by assuring students that there will be no adverse consequences to not participating.
3. Make every effort to design methods that enrich students' educational experiences rather than detract from them.
4. Disseminate the results in ways that protect student identity while also maximizing the benefit of the study for practice.

REFERENCES

Adhikari, A., and D. Nolan. 2002. "But What Good Came of It at Last? How to Assess the Value of Undergraduate Research." *Notices of the American Mathematical Society* 49 (10): pp. 1252–7.

American Association for the Advancement of Science. 2012. *Describing and Measuring Undergraduate STEM Teaching Practices*. Washington: AAAS. Accessed September 2, 2017. http://ccliconference.org/files/2013/11/Measuring-STEM-Teaching-Practices.pdf.

American Statistical Association. 2007. *Using Statistics Effectively in Mathematics Education Research: A Report from a Series of Workshops Organized by the American Statistical Association with Funding from the National Science Foundation*. Accessed September 2, 2017. http://www.amstat.org/education/pdfs/UsingStatisticsEffectivelyinMathEdResearch.pdf.

Anderson, L., and D. A. Krathwohl. 2001. *Taxonomy for Learning, Teaching and Assessing: A Revision of Bloom's Taxonomy of Educational Objectives*. New York: Longman.

Bass, R., and B. Eynon. 2009. "Capturing the Visible Evidence of Invisible Learning (Introduction and Synthesis of Findings)." In *The Difference that Inquiry Makes: A Collaborative Case Study of Technology and Learning, from the Visible Knowledge Project*, edited by R. Bass and B. Eynon. Washington: Academic Commons. Accessed September 2, 2017. https://blogs.commons.georgetown.edu/vkp/files/2009/03/bass-revised-2.pdf.

Berliner, D. 2002. "Educational Research: The Hardest Science of All." *Educational Researcher* 31 (8): pp. 18–20.

Bowen, C. 1994. "Think-Aloud Methods in Chemistry Education: Understanding Student Thinking." *Journal of Chemical Education* 71 (3): pp. 184–90. doi: 10.1021/ed071p184.

Bruff, D. 2015. "Conceptual or Computational? Making Sense of Reading Questions in an Inverted Statistics Course." In *Doing the Scholarship of Teaching and Learning in Mathematics*, edited by J. Dewar and C. Bennett, pp. 127–36. Washington: Mathematical Association of America.

Cohen, J. 1988. *Statistical Power Analysis for the Behavioral Sciences*, 2nd ed. Mahwah: Lawrence Erlbaum Publishers.

Cronbach, L. 1982. "Prudent Aspirations for Social Inquiry." In *The Social Sciences: Their Nature and Uses*, edited by W. H. Kruskal, pp. 61–81. Chicago: University of Chicago Press.

Cook, T. D., and M. R. Payne. 2002. "Objecting to the Objections to Using Random Assignment in Educational Research." In *Evidence Matters: Randomized Trials in Education Research*, edited by F. Mosteller and R. Boruch, pp. 150–78. Washington: Brookings Institution Press.

Dewar, J. 2006. "Increasing Math Majors' Success and Confidence Though Problem Solving and Writing." In *Educational Transformations: The Influences of Stephen I. Brown*, edited by F. Rosamond and L. Copes, pp. 101–21. Bloomington: AuthorHouse.

Dewar, J., S. Larson, and T. Zachariah. 2011. "Group Projects and Civic Engagement in a Quantitative Literacy Course." *PRIMUS: Problems, Resources, and Issues in Mathematics Undergraduate Studies 21* (7): pp. 606–37.

Fedoruk, L. 2017. *Ethics in the Scholarship of Teaching and Learning: Key Principles and Strategies for Ethical Practice.* Taylor Institute for Teaching and Learning Guide Series. Calgary: Taylor Institute for Teaching and Learning at the University of Calgary. Accessed September 2, 2017. http://www.ucalgary.ca/taylorinstitute/guides/ethics-scholarship-teaching-and-learning.

Gieger, L., J. Nardo, K. Schmeichel, and L. Zinner. 2015. "A Quantitative and Qualitative Comparison of Homework Structures in a Multivariable Calculus Class." In *Doing the Scholarship of Teaching and Learning in Mathematics*, edited by J. Dewar and C. Bennett, pp. 67–76. Washington: Mathematical Association of America.

Grauerholz, L., and E. Main. 2013. "Fallacies of SoTL: Rethinking How We Conduct Our Research." In *The Scholarship of Teaching and Learning In and Across the Disciplines*, edited by K. McKinney, pp. 152–68. Bloomington: Indiana University Press.

Hake, R. 1998. "Interactive-Engagement vs. Traditional Methods: A Six-Thousand-Student Survey of Mechanics Test Data for Introductory Physics Courses." *American Journal of Physics 66*: pp. 64–74.

Healey, R. L., T. Bass, J. Caulfield, A. Hoffman, M. K. McGinn, J. Miller-Young, and M. Haigh. 2013. "Being Ethically Minded: Practising the Scholarship of Teaching and Learning in an Ethical Manner." *Teaching and Learning Inquiry: The ISSOTL Journal 1* (2): pp. 23–33. doi: 10.20343/teachlearninqu.1.2.23.

Holcomb, J. 2002. "The Ethics of Comparison: A Statistician Wrestles with the Orthodoxy of a Control Group." In *Ethics of Inquiry: Issues in the Scholarship of Teaching and Learning*, edited by P. Hutchings, pp. 19–26. Menlo Park: The Carnegie Foundation for the Advancement of Teaching.

Hutchings, P., editor. 2002. *Ethics of Inquiry: Issues in the Scholarship of Teaching and Learning.* Menlo Park: The Carnegie Foundation for the Advancement of Teaching.

Institute of Education Sciences and the National Science Foundation. 2013. *Common Guidelines for Education Research and Development.* Accessed September 2, 2017. http://www.nsf.gov/pubs/2013/nsf13126/nsf13126.pdf.

Kagan, J. 2012a. *Psychology's Ghosts: The Crisis in the Profession and the Way Back.* New Haven: Yale University Press.

Kagan, J. 2012b. "Psychology's Missing Contexts". *Chronicle of Higher Education.* April 8. Accessed December 5, 2017. http://chronicle.com/article/Psychologys-Missing-Contexts/131430.

Kaus, C. 2015. "Using SoTL to Assess the Outcomes of Teaching Statistics Through Civic Engagement." In *Doing the Scholarship of Teaching and Learning in Mathematics*, edited by J. Dewar and C. Bennett, pp. 99–106. Washington: Mathematical Association of America.

McKagan, S., E. Sayre, and A. Madsen. 2016. "Effect Size: What Is It and When and How Should I Use It?" PhysPort: American Association of Physics Teachers. Accessed August 30, 2017. https://www.physport.org/recommendations/Entry.cfm?ID=93385.

McKagan, S., E. Sayre, and A. Madsen. 2017. "Normalized Gain: What Is It and When and How Should I Use It?" PhysPort: American Association of Physics Teachers. Accessed August 30, 2017. https://www.physport.org/recommendations/Entry.cfm?ID=93334.

McLean, M., and G. Poole. 2010. "An Introduction to Ethical Considerations for Novices to Research in Teaching and Learning in Canada." *The Canadian Journal for the Scholarship of Teaching and Learning 1* (2). doi: 10.5206/cjsotl-rcacea.2010.2.7.

Meyer, J. H. F., and R. Land. 2003. "Threshold Concepts and Troublesome Knowledge: Linkages to Ways of Thinking and Practicing in the Disciplines." In *Improving Student Learning—Theory and Practice Ten Years On*, edited by C. Rust, pp. 412–24. Oxford: Oxford Centre for Staff and Learning Development.

Mount Royal University Human Research Ethics Board. 2012. *Ethical Considerations for Dual-Role Research: Conducting Research with Students in your own Classroom*. Accessed September 2, 2017. http://www.mtroyal.ca/cs/groups/public/documents/pdf/dualroleresearchers.pdf.

National Research Council. 2002. *Scientific Research in Education*. Washington: The National Academies Press. doi: 10.17226/10236.

Pace, D., and J. Middendorf, editors. 2004. *Decoding the Disciplines: Helping Students Learn Disciplinary Ways of Thinking*, New Directions for Teaching and Learning: No. 98. San Francisco: Jossey-Bass.

Pace, D. 2017. *The Decoding the Disciplines Paradigm: Seven Steps to Increased Learning*. Bloomington: Indiana University Press.

Panel on Research Ethics. 2014. *Tri-Council Policy Statement: Ethical Conduct for Research Involving Humans*. Accessed August 31, 2017. http://www.pre.ethics.gc.ca/pdf/eng/tcps2-2014/TCPS_2_FINAL_Web.pdf.

Poole, G. 2013. "Square One: What Is Research?" In *The Scholarship of Teaching and Learning In and Across the Disciplines*, edited by K. McKinney, pp. 135–51. Bloomington: Indiana University Press.

Ross, P. M., C. E. Taylor, C. Hughes, M. Kofod, N. Whitaker, L. Lutze-Mann, and V. Tzioumis. 2010. "Threshold Concepts: Challenging the Culture of Teaching and Learning Biology." In *Threshold Concepts: From Theory to Practice*, edited by J. H. F. Meyer, R. Land, and C. Baillie, pp. 165–78. Rotterdam: Sense Publishers.

Shulman, L. 2013. "Situated Studies of Teaching and Learning: The New Mainstream." Keynote address at the 10th Annual Conference of the International Society for the Scholarship of Teaching and Learning, Raleigh, NC, October 3. Accessed August 31, 2017. https://www.youtube.com/watch?v=bhvwLW-5zMM.

Stewart, J., and G. Stewart. 2010. "Correcting the Normalized Gain for Guessing." *The Physics Teacher, 48* (3): pp. 194–6. doi: 10.1119/1.3317458.

Suter, L., and J. Frechtling. 2000. *Guiding Principles for Mathematics and Science Education Research Methods: Report of a Workshop*. Washington: National Science Foundation. Accessed August 31, 2017. http://www.nsf.gov/pubs/2000/nsf00113/nsf00113.pdf.

Talanquer, V. 2015. "Threshold Concepts in Chemistry: The Critical Role of Implicit Schemas." *Journal of Chemical Education 92* (1): pp. 3–8. doi: 10.1021/ed500679.

Trafimow, D., and M. Marks. 2015. "Editorial." *Basic and Applied Social Psychology 37* (1): pp. 1–2. doi: 10.1080/01973533.2015.1012991.

U. S. Department of Health and Human Services. 2017. *International Compilation of Human Research Standards*. Accessed August 31, 2017. https://www.hhs.gov/ohrp/sites/default/files/international-compilation-of-human-research-standards-2017.pdf.

Wasserstein, R. L., and N. A. Lazar. 2016. "The ASA's Statement on p-Values: Context, Process, and Purpose." *The American Statistician 70* (2): pp. 129–33. doi: 10.1080/00031305.2016.1154108.

Zolan, M., S. Strome, and R. Innes. 2004. "Decoding Genetics and Molecular Biology: Sharing the Movies in Our Head." In *Decoding the Disciplines: Helping Students Learn Disciplinary Ways of Thinking, New Directions for Teaching and Learning*: No. 98, edited by D. Pace and J. Middendorf, pp. 23–32. San Francisco: Jossey-Bass. doi: 10.1002/tl.144.

CHAPTER 4

Gathering Evidence

The Basics

Introduction

The scholarship of teaching and learning (SoTL) involves the systematic investigation of a question we have about student learning. Once we formulate a researchable question, we have to gather and then analyze evidence. This chapter examines basic considerations of research design related to evidence, such as whether, and how, to gather quantitative data, qualitative data, or both.

Triangulating Data

A SoTL researcher should develop a plan for systematically collecting multiple types of evidence. A diversity of evidence can help the researcher to form a convincing picture of student learning (Wiggins 1998). This approach is called **triangulation**. According to Webb et al. (as cited in National Research Council 2002, 64): "When a hypothesis can survive the confrontation of a series of complementary methods of testing, it contains a degree of validity unattainable by one tested within the more constricted framework of a single method." In other words, claims or explanations supported by several types of evidence—for example, student work samples, interviews, and retention rates—are considered to be more accurate. This will be an asset if the work is submitted for publication in a peer-reviewed journal.

The Scholarship of Teaching and Learning, Jacqueline M. Dewar, Curtis D. Bennett, and Matthew A. Fisher.
© Oxford University Press, 2018. Published 2018 by Oxford University Press

We next shift our attention to considerations of the two types of data we might collect in a SoTL inquiry: quantitative and qualitative. A study that relies on both types is often referred to as a **mixed method research design**. In writing about its use in chemistry education research, Towns (2008) observed: "Mixed methods designs allow researchers to use both qualitative and quantitative methods in the same study in order to balance the inherent strengths and weaknesses of each research methodology. The sequential or concurrent engagement of both research methodologies can lead to more interpretable and valid outcomes than either approach could provide alone" (135). Towns (2008, 145–6) gave seven examples of mixed methods designs in chemistry education research from the *Journal of Chemical Education* and the *Journal of Research in Science Teaching*. For each article, a brief description of the data collection and analysis or research design was provided.

Quantitative Versus Qualitative Data

It seems easy enough to distinguish between the two types of data. **Quantitative data** are numerical data, such as test scores, grade point averages, Likert scale survey data, or percent completion rates. **Qualitative data** are anything else, that is, non-numerical data. This includes lab reports, computer programs, solutions to problems, proofs, and reflective writing by students, drawings or diagrams, text from interviews, video recordings of students solving problems in groups, or characteristics such as gender. Usually, scientists, engineers, and mathematicians are more familiar with quantitative data. They tend to believe that quantitative data are more scientific, rigorous, or reliable, a view held by many, but not all, social science researchers as well. However, faculty members in disciplines such as anthropology or educational psychology may have a different view. From their perspective, qualitative data have advantages because they provide more in-depth or nuanced understandings of the subject under investigation. They allow the investigator to examine "Why?" or "How?" questions that quantitative data are often ill-suited to address. Recalling Hutchings' (2000) taxonomy of SoTL questions from Chapter 2 (A Taxonomy of SoTL Questions), quantitative data may be a better fit for *What works?* questions while qualitative data more naturally yield insights into *What is?* or *What could be?* questions.

However, Trochim (2006) cautioned that identifying data as quantitative or qualitative creates a false dichotomy. All quantitative data at some point almost certainly involved qualitative judgments. For instance, think about a math anxiety scale. Developing it required numerous decisions: how to define math anxiety, how to phrase potential scale items, how to ensure the items are clearly stated for the intended subjects, what cultural and language constraints might occur, and so on. The investigator who decides to use the scale has to make another set of judgments, such as whether the scale relates to the research question and if it is appropriate for the subjects involved in the study. When the subjects fill out the scale, they will make many more judgments: what the various terms and phrases mean, why the researcher is giving this scale to them, and how much energy and effort they want to expend to complete it.

If this discussion of a math anxiety scale, adapted from Trochim (2006), seems too obscure, consider a typical physics, general chemistry, or introductory biology exam that yields a percentage score for each student. The instructor decided how many and which problems to include, how they would be worded, and how partial credit would be given. The students made their own judgments about what the questions were asking, how much work to show (on questions that were not multiple choice), and how much time and effort to expend. In both examples, what may look like a simple quantitative measure is based on numerous qualitative judgments made by many different individuals.

Conversely, virtually all qualitative data can be transformed, expressed, and analyzed numerically. Much like quantitative data, qualitative data can be categorized, counted, or turned into percentages, and used to compare populations. We will describe how to do this with qualitative data in Chapter 7 (Coding Data).

Potential data sources for SoTL studies are quite varied. We begin by considering those sources, both quantitative and qualitative, likely to be familiar to most science, technology, engineering, or mathematics (STEM) faculty members. These include items that students would typically generate as they do their work in a course, as well as institutional research data and survey data.

Familiar Sources of Evidence

Science, engineering, and mathematics instructors not trained in disciplinary-based education research are typically most familiar and comfortable with gathering and interpreting test scores, grade point averages, time-on-task, or percent completion rates. Table 4.1 displays many kinds of numerical data that could be gathered for a SoTL study.

Table 4.1 *Examples of familiar quantitative (numerical) data that might be used in a SoTL study*

Familiar quantitative (numerical) data

Scores
- Pre- and post-tests
- Course work and homework assignments
- Quizzes, mid-terms, or final exams
- Lab reports, papers, and projects
- Standardized scales, inventories, and tests (see Appendix IV for a list of examples)

Survey results
- Surveys of attitudes, beliefs, or satisfaction, often using a Likert scale from "Strongly Disagree" to "Strongly Agree"
- Student ratings of teaching

Frequency counts or percentages
- Multiple-choice test item responses
- Course completion rates
- Participation in class, on discussion boards, etc.
- Online homework system usage
- Office visits

Measures of time use
- Time spent online accessing homework systems or other resources

Institutional research data
- Academic transcript data (e.g., grades, GPA, admission or placement test scores)
- Retention data (e.g., in course, program, major, or institution)
- Enrollment in follow-up courses
- Student demographics

While most of these data items seem straightforward to collect and interpret, complications can arise even with these very familiar forms of numerical data. As Light et al. (1990) put it: "You can't fix by analysis what you bungle by design" (viii). For example, in collecting time-on-task data from online homework systems, how long a student user has a webpage open may be misleading as an indicator of time-on-task. In Chapter 3 (Aligning the Assessment Measure with What is Being Studied) we highlighted another pitfall, namely, collecting and trying to interpret data that is not well-aligned with the factors or attributes under investigation.

In addition to the familiar sources of quantitative data just described, STEM instructors would have experience with the non-numerical data that result from lab reports, research papers, computer programs, design diagrams, or models

Table 4.2 *Examples of familiar qualitative (non-numerical) data that might be used in a SoTL study*

Familiar qualitative (non-numerical) data

Solutions to
- Pre- and post-tests
- Exam or quiz questions
- Homework problems

Writing assignments
- Projects or papers
- Portfolios
- Reflective writing or journals
- Minute papers and other classroom assessments (see Angelo and Cross 1993; Barkley and Major 2016)

Discipline-specific assignments
- Lab reports
- Computer programs
- Designs for scientific experiments
- Drawings (e.g., of chemical bonds or molecules)

Open-ended responses on surveys

(see Table 4.2 for additional examples). While most instructors have experience assigning, tallying, or grading some of the types of data listed in Table 4.2, they are probably far less familiar with methods for analyzing these data to answer a SoTL question. Such methods include rubrics and content analysis, also called coding. These topics are the focus of Chapter 7.

Instructors often determine a numerical grade for solutions, papers, or projects without applying a rubric. However, when these qualitative data are used as evidence of student learning in a SoTL study, a rubric is often helpful. It can yield numerical data for multiple dimensions or characteristics of the work, making possible a more detailed assessment of learning than a single grade on a problem or a paper.

Another common approach to analyzing written work obtained from portfolios, journals, or open-ended survey questions as part of a SoTL project is to code or categorize the text for commonalities. These might be themes found in students' reflective writing or common mistakes, false starts, or misunderstandings in proofs. These two methods of analyzing qualitative data—rubrics and coding data—are described in more detail in Chapter 7 (Analyzing Evidence).

Assignment Design

The design of assignments, specifically, the actions required to complete them, can have a profound impact on the learning that takes place (Weimer 2015) as well as affect the quality of evidence we might obtain from the students' work. Weimer cited a study of sociology students (Foster 2015) and how the audience for their writing (private journal or public blog) changed the learning outcome. In his book *Creating Significant Learning Experiences*, Fink (2013) emphasized the importance of aligning learning activities and assessments with the learning outcomes at the course design level. Berheide (2007) argued for using data that students produce in their capstone courses for assessment purposes because this approach "is simply easier than most other assessment options" and "provides better measures of student learning" (27). Her rationale applies equally well to using student work as evidence in SoTL studies.

Both the students and the instructor benefit when a SoTL study is designed so that the coursework students produce can also serve as evidence. The SoTL study is not adding to the students' workload and the instructor does not need to design a separate instrument to gather that evidence. Of course, the learning activity and the assessment of it must align with the desired student learning outcome and with the SoTL research question. We examine some assessment techniques that might be used to do this.

Many STEM faculty may be familiar with **one minute** or **muddiest point papers**—short written responses to a question posed by the instructor to gain student feedback as opposed to evaluate student learning. These are examples of what have become known as **classroom assessment techniques** (CATs), first documented by Angelo and Cross (1993) in their volume of the same name. CATs are formative assessments designed to give faculty insight into student comprehension of course content or their ability to analyze or synthesize concepts. The assessments are usually non-graded and almost always anonymous. Angelo and Cross arranged the 50 techniques in their book into three broad categories according to their purpose: techniques for assessing course-related knowledge and skills; techniques for assessing learner attitudes, values, and self-awareness; and techniques for assessing learner reactions to instruction. Each category was further subdivided. This organization makes it easy to find the CATs that might be appropriate for a given research question. In addition, for each of the 50 techniques the book gives a step-by-step procedure and an example, provides estimates (low, medium, high) of how much time and energy is required (to prepare the CAT, to respond to the CAT, and to analyze the data obtained), discusses how to turn the data collected into useful information, discusses pros and cons of using the CAT, and offers very specific advice for using it and what to expect as a result. This level of detailed information facilitates the use of CATs.

More recently, Barkley and Major (2016) have revisited classroom assessment techniques along with collaborative learning techniques and student engagement techniques to compile what they refer to as **learning assessment techniques.** These techniques result in the creation of some artifact of student learning that can be used as evidence in a SoTL study. Barkley and Major also stressed the importance of first attending to the desired learning outcome and reason for assessing it before choosing how to perform the assessment.

Bean (2011), Felder and Brent (2016), and Hodges (2015) contain additional resources for designing assignments that can also serve as evidence in a SoTL study. Bean (2011) will be of particular interest to STEM faculty looking to design assignments focused on writing or reading; it includes an entire chapter on designing tasks that promote active thinking and learning in the context of both formal and informal or exploratory writing activities.

> **Pause: Gathering Evidence—Questions to Consider**
> - Which kind of data—quantitative, qualitative, or both—is a good match for the type of research question (*What works? What is? What could be?*) that I have?
> - What data are already readily available?
> - Do individual questions or items on my tests, surveys, interviews, etc. align well with my research question?
> - How will I triangulate my data?
> - Can I design course assignments in such a way that they will also generate evidence that could be used in my study?
> - Have I thought about using classroom assessment techniques to generate evidence for my study?

REFERENCES

Angelo, T. A., and K. P. Cross. 1993. *Classroom Assessment Techniques: A Handbook for College Teachers*, 2nd ed. San Francisco: Jossey-Bass.

Barkley, E. F., and C. H. Major. 2016. *Learning Assessment Techniques: A Handbook for College Faculty.* San Francisco: Jossey-Bass.

Bean. J. C. 2011. *Engaging Ideas: The Professor's Guide to Integrating Writing, Critical Thinking, and Active Learning in the Classroom*, 2nd ed. San Francisco: Jossey-Bass.

Berheide, C. W. 2007. "Doing Less Work, Collecting Better Data: Using Capstone Courses to Assess Learning." *Peer Review 9* (2): pp. 27–30.

Felder, R., and R. Brent. 2016. *Teaching and Learning STEM: A Practical Guide*. San Francisco: Jossey-Bass.

Fink, D. 2013. *Creating Significant Learning Experiences*, 2nd ed. San Francisco: Jossey-Bass.

Foster, D. 2015. "Private Journals versus Public Blogs: The Impact of Peer Readership on Low-Stakes Reflective Writing." *Teaching Sociology 43* (2): pp. 104–14.

Hodges, L. 2015. *Teaching Undergraduate Science: A Guide to Overcoming Obstacles to Student Learning*. Sterling: Stylus Publications.

Hutchings, P., editor. 2000. *Opening Lines: Approaches to the Scholarship of Teaching and Learning*. Menlo Park: The Carnegie Foundation for the Advancement of Teaching.

Light, R., J. Singer, and J. Willett. 1990. *By Design: Planning Research on Higher Education*. Cambridge: Harvard University Press.

National Research Council. 2002. *Scientific Research in Education*. Washington: The National Academies Press. doi: 10.17226/10236.

Towns, M. H. 2008. "Mixed Methods Designs in Chemical Education Research." In *Nuts and Bolts of Chemical Education Research*, edited by D. Bunce and R. Cole, pp. 135–48. Washington: American Chemical Society.

Trochim, W. 2006. "Types of Data." *Research Methods Knowledge Base*. Updated October 20. http://www.socialresearchmethods.net/kb/datatype.php.

Weimer, M. 2015. "How Assignment Design Shapes Student Learning." Teaching Professor Blog. Accessed April 8, 2017. http://www.facultyfocus.com/articles/teaching-professor-blog/how-assignment-design-shapes-student-learning.

Wiggins, G. 1998. *Educative Assessment*. San Francisco: Jossey-Bass.

CHAPTER 5

Evidence

From Surveys

Introduction

This chapter discusses surveys, beginning with the kind most people have experienced taking, one that inquires about agreement, satisfaction, or participation. It usually generates numerical data but may also ask free-response questions that produce qualitative data. The second type of survey that we consider, the **knowledge survey**, produces only quantitative data. It is intended to measure students' confidence in their knowledge of disciplinary content. Science, technology, engineering, and mathematics (STEM) instructors may or may not be familiar with knowledge surveys.

Surveys

Because surveys are familiar instruments and online survey systems can make it easy to conduct a survey, many instructors interested in studying student learning or motivation decide to collect survey data. There are many ways to set up the response options for survey items, but a **Likert scale** with responses ranging from Strongly Disagree to Strongly Agree is often used. These responses can be converted to a numerical scale ranging from 1 to 5 if a neutral option is provided, and 1 to 4 if it is not, though more or fewer gradations in the scale are also common. With sufficiently many student responses, the pre- and post-results can be compared and tested for statistical significance.

The Scholarship of Teaching and Learning, Jacqueline M. Dewar, Curtis D. Bennett, and Matthew A. Fisher.
© Oxford University Press, 2018. Published 2018 by Oxford University Press

As an indication of how popular surveys are as a method of data collection in SoTL studies, of the 15 investigations into mathematical learning included in Dewar and Bennett (2015), ten employed surveys. One investigator used a professionally developed survey (Holcomb, 2015) and another (Kaus, 2015) revised one to suit her purposes. The remainder chose to design their own surveys, some from scratch and others after looking at various existing surveys. (See Appendix IV for a list of survey instruments that might be useful in conducting scholarship of teaching and learning (SoTL) studies in STEM fields.) Many noted concerns and pitfalls that they encountered either in the design or in the administration of their surveys (for example, see Gieger et al. 2015; Mellor 2015). Problems with survey completion rates were reported by several investigators (Holcomb 2015; Kaus 2015; Mellor 2015), along with suggestions for improving response rates (see Improving Response Rates, this chapter). Szydlik (2015) provided a lengthy discussion of the steps he took to design, pilot, and validate his survey instrument. In one study, unexpected survey results suggested the need to collect new or more nuanced data (O'Hanlon et al. 2015).

Essential Steps in Designing a Survey

Because the use of self-constructed surveys is so common in SoTL studies, we present a summary of critical steps in designing surveys. They were drawn from online resources (https://academics.lmu.edu/spee/officeofinstitutionalresearch/surveysevaluation/surveydesign) of the Office of Institutional Research at Loyola Marymount University, which cited Suskie (1996). We recommend seeing if there is a similar office or person on campus that supports the administration of surveys by faculty members doing assessment or research studies that could provide additional information, support, or access to survey software.

Step 1. Identify the objectives and the target population for the survey.

- Before designing the survey, clearly define its purpose. Exactly what information is the survey intended to capture?
- Think carefully about the target population, that is, who will take the survey? The characteristics of the population (e.g., students enrolled in certain courses, students using online homework systems, faculty members, future teachers, etc.) may limit the type of information that can be collected from them. Consider whether there are cultural or other factors that may influence their responses to the questions.

- Think how the information will be used for the study. To make comparisons between different groups of people, include questions that will enable appropriate grouping of respondents.

Step 2. Construct well-written items.

The quality of the data gathered will depend on the quality of the items. Good items have the following characteristics:

- Each item is aligned with one or more of the objectives.
- The items are stated in the simplest possible language. Any complex terms used in the survey are defined.
- The wording is specific and precise so as to avoid confusion or misunderstandings.
- Each item is concise and to the point. Long questions take time to read and are difficult to understand.
- The items contain few, if any, negative words and they avoid double negatives. Negative statements tend to confuse survey takers.
- The items are not "double-barreled." A double-barreled item inquires about two or more parameters but allows only a single response. Asking students in a single item whether the online homework system provided *timely and useful feedback* is an example of a double-barreled question.
- The items are not biased nor do they lead the respondent toward a particular answer or perspective. An example of an item that suggests a particular viewpoint (when presented for agreement or disagreement) is the following: *The next workshop should allow time for people to interact in small groups. It was sometimes difficult to have a voice in such a large group.*

Step 3. Choose the format of the response options.

A wide variety of response options exists for survey items. Here are the most commonly used response options along with special considerations for each.

- Yes/No options. Be careful using this simple response option as it allows for no grey area in between. The yes/no option is useful for contingency questions to determine if subsequent items apply or should be skipped.
- Multiple-choice options. These provide a fixed set of answers to choose from and can be designed to allow the selection of one or multiple responses. Response options should be mutually exclusive. If the options are not distinct, respondents will not know which to select. Capture all possible responses while using the smallest number of categories

possible. Including an "Other" category followed by a request to please "Explain" or "Specify" will capture overlooked options.
- Likert scales. These ask respondents to rate their preferences, attitudes, or subjective feelings along a scale, such as from "Strongly Disagree" to "Strongly Agree." Each option in the scale should have a label and a decision must be made whether to include a neutral option in the center of the scale, as well as what direction the scale should run. For each topic or dimension being surveyed, framing some items positively and some negatively makes it possible to check that respondents are reading carefully. Some studies have shown that the order of the response options can affect the results, finding that a scale going agree to disagree from left to right produces greater agreement, than a scale going from disagree to agree (Hartley 2014; Hartley and Betts 2010).
- Alternative responses. Another judgment to be made is whether to offer a response such as Don't Know, Decline to State, Not Applicable, or Other as a way to allow survey-takers to avoid giving a direct answer. Being forced to answer, especially if the options fail to include a correct or truthful response, can be frustrating to respondents, lead them to abandon the survey, or lower the accuracy of the data obtained.
- Open-ended responses. Open-ended responses allow survey-takers to answer in their own words. This type of response provides a rich source of information, but requires more time to analyze. The textual data will have to be coded for common themes, a process described in Chapter 7 (Coding Data).

For any format, order all response options logically and be consistent with the ordering for similar questions in the survey.

Step 4. Organize the survey with care.

A well-organized survey can improve the response rate. Here are some suggestions.

- Begin with an introduction that explains the purpose of the survey and how the person's participation will help. Address privacy concerns by stating that participation is voluntary, and responses will be confidential. If the survey will also be anonymous, inform respondents of that. State who is conducting the survey and whom to contact with questions or concerns.
- Group items by topic and put groupings in a logical order.

- Put demographic questions at the end of the survey and give respondents the option of "decline to state."
- Make it as short as possible to avoid survey "fatigue," which can result in loss of or poorer quality data.
- Close with a thank you and repeat the contact information for any questions or concerns.

Step 5. Pilot test the survey and make any necessary changes.

Pilot testing is important because it is all too easy to write a confusing item unintentionally or to have an error in the logic of an online survey.

- Recruit a small group of people to take the survey and give feedback on the clarity of the directions and the items and how long the survey took.
- Have the test group include some individuals who are similar to the target population.
- Review the completed responses for unexpected answers or inconsistencies.
- If the pilot prompted a large number of changes to the survey, consider conducting another pilot test on the revised survey.

Reliability and Validity

We want the data we gather from our survey to be useful in answering our question. Thus the survey instrument must be **reliable**, that is, yield consistent results, and **valid**, meaning it measures what it is intended to measure. An in-depth discussion of determining reliability and validity for surveys and other assessment measures is beyond the scope of this book. We found the discussion of reliability and validity in Bretz (2014) and the examples given there from chemistry education research quite accessible.

Rather than surveying students individually, some studies have small groups of students confer and then fill out a survey as a group. Moy et al. (2010) took this approach, noting that Finney (1981) and Poggio et al. (1987) recommended using student panels that discuss and come to consensus to overcome the intrinsic unreliability of retrospective self-assessment of learning gains. Moy et al. also asked in two different ways (first with a Likert scale and then by prioritizing the resources) how seven different resources—lecture, textbook, problem sets, literature papers, Wikipedia, working alone, and working in groups—influenced the attainment of five different learning goals. Since the second question was given with a reversed scale, this allowed the researchers to check the reliability of the answers to the first. (See their article for more details.)

Improving Response Rates

Many new SoTL investigators who try to gather survey data are unaware of how difficult it can be to get students to respond to surveys, let alone to capture a sufficient number of matched pre- and post-surveys in order to make statistical comparisons. The reasons for low **response rates** are many. The number of survey requests students receive is increasing, both from their institution (for example: surveys from housing, student government, and program assessment efforts, as well as national surveys of student engagement), and from non-academic sources. It is easy for students to ignore or overlook survey requests that come by email, especially if the requests come to their "school" email address, when they often rely on their "personal" email account or other forms of social media to communicate with others. Other factors contributing to non-response include lack of interest in the topic or understanding why it should matter to them, the request arriving at an inconvenient or busy time (like the end of the semester), and poor survey design that leads them to quit before finishing the survey due to irritation or frustration. Low response rates are a particular problem if the researcher hopes to obtain generalizable results because those who do respond may not be representative of the population of interest. In all studies, but especially in those where making generalizations is a goal, it is a good idea to have some survey questions at the end to gather demographic or other data to determine how well the respondents represent the entire group. A discussion of how characteristic the sample is should be included in any published results. Some suggestions for improving response rates follow.

Knowing the target population and their circumstances is important in order to address many of the non-response factors. There are many reasons to use an online survey, but one advantage of surveys administered on paper in class is a higher response rate (Nulty 2008). A major disadvantage of paper surveys is the data has to be analyzed by hand or entered into a program and some way of matching pre- and post-responses must be devised unless students are asked to identify themselves. Both paper and online surveys administered in class require setting aside time, and extra effort is needed to get data from those not in attendance on the day of the survey. That the views of those in attendance may differ from those absent is another concern.

Try to schedule the survey wisely; some schools keep a calendar of surveys that faculty can consult. The end of the semester is always a difficult time, especially for seniors, and once the semester ends, the response rates will drop tremendously. Institutions often incentivize responses by offering prizes to respondents awarded through a drawing. If institutional policies allow it, individual faculty members can offer a small grade incentive, either to individuals or to the whole class if a certain response rate is achieved (Dommeyer et al. 2004).

Fink (2013) offers these additional suggestions. Make sure the survey is well designed (the questions are clear and answer options allow all students to answer truthfully) and short (at most ten minutes to take it). Provide a progress bar whenever that feature is available in the online survey tool. Explain why the survey should matter to the students, and who will benefit; offer to send an aggregate report, if appropriate. If none of the survey questions allow for a free response, students may not feel their opinions are valued, so consider including one or more free response items. Address privacy concerns by clarifying who will have access to the data, whether the responses will be confidential (no identifying characteristics are reported) or anonymous (no names or identities are collected), and whom to contact if participants have any questions about the survey. Provide sufficient time to respond and send one or two reminders to nonresponders of online surveys.

What is an Acceptable Response Rate?

The answer depends, in part, on the purpose for the survey. Is the purpose to describe the group's characteristics, to test for pre-post differences, or to look for differences in subgroups? If the research design includes testing for significant differences, then sample size is an important consideration. How well the respondents represent the entire group under study is also a key concern. Because an inherent bias may result in a low response rate, low response rates can degrade the credibility of results more than a small sample size. While there is no agreed-upon standard to be found in the literature, some journals are very explicit about their requirements for publication of survey research. For example, the *American Journal of Pharmaceutical Education Research* set 60 percent as the goal for the response rate for most research, with a higher level (≥ 80 percent) if the research is intended to represent all schools and colleges of pharmacy (Fincham 2008). In reporting results, explain how the survey was prepared and administered, what the response rate was, and how representative the group of respondents was.

Adhikari and Nolan (2002), two statisticians, offered additional pointers on the construction, administration, and analysis of surveys based on their experience evaluating summer undergraduate research programs at Mills College and the University of California, Berkeley. Draugalis et al. (2008) provided ten guiding questions and recommendations for what they considered to be best practices for research reports that utilize surveys.

Examples of STEM Education Studies that Used Surveys

A study of the impact that the model minority stereotype has on Asian-American engineering students (Trytten et al. 2012) used a mixed method approach that included analysis of academic transcript data, surveys, and semi-structured interviews. The surveys included questions about racial and ethnic background, attitudes about engineering, and confidence in various engineering-related skills. The non-demographic questions were adapted from the Pittsburgh Engineering Attitudes Survey (Besterfield-Sacre et al. 1997).

Smith et al. (2015) investigated whether and how stereotype threat concerns might influence women's science identification. Their quantitative study relied entirely on survey data from 388 women enrolled in introductory physics (male-dominated) and biology (female-dominated) undergraduate laboratory classes at three universities. Their results suggested that women's identification with and persistence in physics may be enhanced when its altruistic value, that is, how much it involves working with and helping other people, is revealed.

> **Pause: Surveys—Questions to Consider**
> - Is a survey a good match for the type of research question (*What works? What is? What could be?*) that I have?
> - Will I use an already developed survey or design one of my own?
> - If I design my own, have I followed the recommendations for good survey design, and have I addressed concerns about validity and reliability?
> - What steps will I take to achieve an acceptable response rate?
> - How will I determine how closely those who respond match the target population?

Knowledge Surveys

We turn our attention to a different type of survey that may not be so familiar. A **knowledge survey** is a tool designed to assess changes in students' confidence about their knowledge of disciplinary content. Nuhfer and Knipp (2003) first described this tool in *To Improve the Academy*, an annual publication for faculty developers, the people who direct campus teaching and learning centers. Knowledge surveys are useful for course development, revision, or improvement efforts. They have also been employed to document success or, more importantly, failure in student learning in courses or programs. For example, the Loyola Marymount University Mathematics Department successfully employed knowledge surveys in an assessment of its engineering calculus sequence (Dewar 2010). However, their use as a valid measure of student learning is not without controversy, as we shall discuss (see Accuracy of Knowledge Surveys, in this chapter). Nevertheless, we believe they have their place in SoTL work, provided that the researcher uses multiple measures to triangulate the data gathered from a knowledge survey. Their use offers additional benefits to the instructor, such as being a way to examine student background knowledge and to study course content sequencing and timing. They provide students with a preview of course content, a way of monitoring their progress throughout the semester, and a study guide for a final exam.

How do knowledge surveys work? Knowledge surveys are given to students at the beginning and the end of a course or a unit. They contain questions about every content item or learning objective, anything that students are supposed to know or be able to do at the end of the course or unit. For a typical first semester calculus course, one would expect to see questions about: computing a limit, a derivative, and an area under a curve; explaining the relationship between continuity and differentiability; finding the maximum value of a function; finding the interval(s) where a function is increasing or decreasing; using the derivative as a linear approximation to a function; interpreting a differential equation geometrically via a slope field; and setting up and solving problems on a variety of topics such as velocity, acceleration, related rates, Riemann sums, and the Fundamental Theorem of Calculus. Knowledge survey questions are phrased like quiz or test questions and are presented in the same chronological order as the content appears in the course (or unit or program). The unusual aspect is that students do not actually answer the questions or work the problems. Instead they rate their confidence to answer each question correctly on a scale of 1 to 3, where 3 means they are confident they can answer correctly right now; 2 means they could get it 50 percent correct—or if given a chance to look something up, and they know just what to look up, they are sure they would then get the answer 100 percent correct; and 1 means they have no confidence that they could answer the question, even if allowed to look something up.

Knowledge surveys were one of several measures of student learning in a quantitative literacy (QL) course that incorporated a civic engagement component (Dewar et al. 2011). Table 5.1 contains knowledge survey questions for that QL course.

Table 5.1 A knowledge survey for a quantitative literacy course (based on Dewar et al. 2011)

Knowledge Survey Question	Confidence Rating
1. Determine the number of square inches in 12 square yards.	1 2 3
2. The speed limit posted as you leave Tecate, Mexico is 50 km/hr. Find the corresponding limit in miles per hour.	1 2 3
3. A department store advertised an 80% off sale on fall apparel. The ad also contained a coupon for an extra 15% off to be applied to the reduced price of any sale or clearance purchase. Find the final price of a $150 suit (ignore tax).	1 2 3
4. Is it possible to go on a diet and decrease your calorie intake by 125%? Explain why or why not.	1 2 3
5. Describe any likely sources of random or systematic errors in measuring the numbers of popped kernels in "large" boxes of popcorn at a movie theater.	1 2 3
6. Show that you understand the difference between absolute error and relative error by giving two examples: one where the absolute error is large but the relative error small, and the other where the absolute error is small but the relative error is large.	1 2 3
7. Use the appropriate rounding rules to answer the following with the correct precision or number of significant digits: find the total weight of a 50 kg bag of sand and a 1.25 kg box of nails.	1 2 3
8. On what basis, if any, would you question the following statistic: the population of the United States in 1860 was 31,443,321?	1 2 3
9. Suppose you win a $100,000 raffle. You wisely invest half of it in a savings account that pays interest with an APR of 5% that will be compounded quarterly. Find how much you will have in the account in 10 years.	1 2 3
10. What does the Annual Percentage Yield (APY) of an investment measure?	1 2 3
11. Suppose you want to start a savings program for a down payment on a house. In 10 years you would like to have $125,000. Your financial advisor can find an account with an APR of 7% that will be compounded monthly. Find how much you will have to deposit into the account per month in order to have the $125,000 in 10 years.	1 2 3

continued

Table 5.1 *Continued*

Knowledge Survey Question	Confidence Rating
12. You wish to buy a new car and can afford to pay at most $400 per month in car payments. If you can obtain a 4-year loan with an APR of 3.2% compounded monthly, what is the largest loan principal you can afford to take out? Round your answer to the nearest dollar.	1 2 3
13. You just received a $1000 credit card bill, and your card has an annual interest rate of 18%. Your credit card company uses the unpaid balance method (i.e., charges interest on the unpaid balance) in order to calculate the interest you owe. Suppose you make a $200 payment now, and make no new charges to your credit card in the next month. Find the balance on your next credit card bill a month from now.	1 2 3
14. The United States has a progressive income tax. Explain what that means.	1 2 3
15. Which is more valuable to a taxpayer, a tax deduction or a tax credit? Explain why.	1 2 3
16. Describe at least three misleading perceptual distortions that can arise in graphics.	1 2 3
17. A company has 10 employees, making the following annual salaries: 3 make $20,000, 2 make $30,000, 4 make $50,000, and 1 makes $1,200,000 per year. Explain whether the median or the mean would be a better representation of the "average" salary at the company.	1 2 3
18. Two grocery stores have the same mean time waiting in line, but different standard deviations. In which store would you expect the customer to complain more about the waiting time? Explain.	1 2 3
19. What are quartiles of a distribution and how do we find them?	1 2 3
20. Give a five-number summary and depict it with a boxplot for the following set of data: {2, 5, 3, 4, 4, 6, 7, 5, 2, 10, 8, 4, 15}.	1 2 3
21. The body weights for 6-month-old baby boys are normally distributed with a mean of 17.25 pounds and standard deviation of 2 pounds. Your 6-month-old son Jeremiah weighs 21.25 pounds. Jeremiah weighs more than what percentage of other 6-month-old baby boys?	1 2 3
22. In order to determine how many students at LMU have ever used a fake ID to buy liquor, we survey the students in this class and find 40% of them have done so. We conclude 40% of LMU students have used a fake ID to buy liquor. Discuss possible sources of bias in the sample and comment if the conclusion is justified.	1 2 3
23. A survey of 1,001 randomly selected Americans, age 18 and older, was conducted April 27–30, 2000, by Jobs for the Future, a Boston-based research firm. They found that 94% of Americans agree that "people who work full-time should be able to earn enough to keep their families out of poverty." Explain what is meant by saying the margin of error for this poll at the 95% confidence interval is 3%.	1 2 3

KNOWLEDGE SURVEYS | 79

Knowledge Survey Question	Confidence Rating
24. Formulate the null and alternative hypotheses for a hypothesis test of the following case: A consumer group claims that the amount of preservative added to Krunch-Chip brand of potato chips exceeds the 0.015 mg amount listed on the packages.	1 2 3
25. Describe the two possible outcomes for the hypothesis test in #24.	1 2 3
26. A random sample of Krunch-Chip potato chip bags is found to have a mean of 0.017 mg of preservative per bag. Suppose 0.03 is the probability of obtaining this sample mean when the actual mean is 0.015 mg preservative per bag as the company claims in #24. Does this sample provide evidence for rejecting the null hypothesis? Explain.	1 2 3

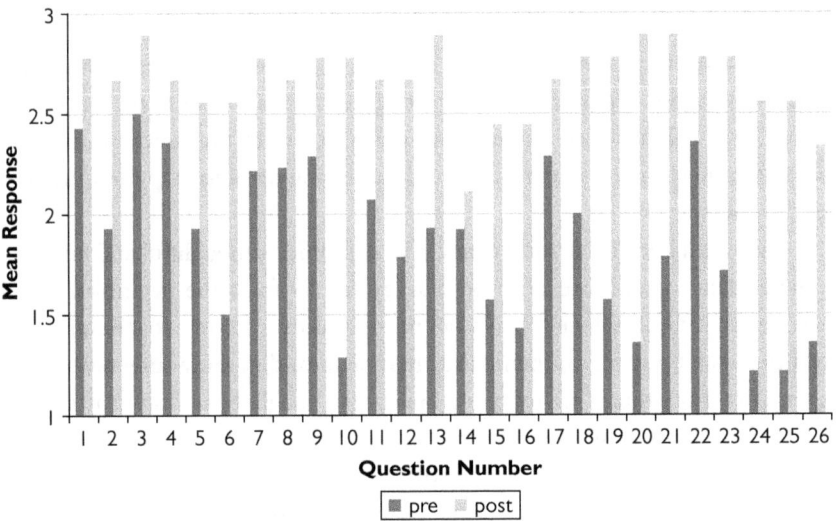

Figure 5.1 Pre- and post-results for the knowledge survey questions given in Table 5.1 when administered in a quantitative literacy course in Spring 2006 (Dewar et al. 2011).

Figure 5.1 provides a double bar graph for the pre- and post-survey results from a Spring 2006 QL course for each question given in Table 5.1.

Several observations emerge almost immediately from the graph in Figure 5.1:

- The mean pre- and post-responses to question #14 indicate a problem with student learning.

- Confidence falls off for the topics at the end of the semester.
- Student confidence was fairly high for some of the items at the beginning of the semester.

We will elaborate on each of these observations after discussing the advantages knowledge surveys hold over pre- and post-tests. Unlike typical pre- and post-tests, knowledge surveys can be administered as take-home work rather than using class time for them. Since rating confidence about doing a problem takes a lot less time than actually carrying out the solution or writing an explanation, course topics can be covered in detail. Many knowledge surveys contain 100 or more questions. The example in Table 5.1 is brief as knowledge surveys go, because a significant amount of the coursework involved students working in groups to apply mathematics to a campus or community issue, and that work was assessed in a different way. Since making confidence judgments takes little of the students' time, the survey can include complex higher order questions (apply, explain, describe, justify, design) as well as basic skills questions (find, calculate). If students are given access to the survey during the semester, they can use it as a study guide, allowing them to map their progress in the course.

Preparing a knowledge survey for a course promotes thoughtful and informed course planning and organization. Survey results from the beginning of the course can indicate which topics students may already know a lot about, and so less time can be spent on them. They may suggest that more time needs to be spent on certain topics. In multi-section courses taught by multiple instructors, a knowledge survey can encourage more consistent coverage of content, especially if the instructors collaborate on constructing it.

We return now to the earlier observations about the graph in Figure 5.1. The small improvement in confidence on item #14—a question about the meaning of a progressive income tax—disappointed the instructor (Dewar). This prompted her to reflect on how she had presented the topic in the course: by lecturing, with little student involvement, and no assigned homework or quiz questions on it. The knowledge survey enabled her to identify a failure of instruction, as well as a failure of learning. She concluded that if being able to explain what a progressive income tax means was an important learning outcome for the QL course, she had to revise her instruction to engage the students with the topic. After a change in the instruction, the knowledge survey could again be used to see if there was any improvement in students' confidence ratings.

The additional observations from the graph (Figure 5.1) that student confidence was fairly high for some of the items at the beginning of the semester, but fell off for topics at the end of the semester, also bear further discussion. Both are common occurrences in knowledge surveys. As Nuhfer and Knipp (2003) observed, students tend to be overconfident on the pre-survey, and typically the post-survey shows lower confidence levels for end-of-semester material. The latter may be due to a rush to cover topics before the end of the semester or students not yet having had much time to study the material. Student overconfidence at the beginning suggests that additional background knowledge checks should be undertaken before skipping or lightening coverage of critical topics.

How hard are knowledge surveys to make up? Anyone who has previously taught the course will probably find most of the needed material in old quizzes and tests or textbook exercises. It is important to check that all learning objectives are addressed and that only material to be taught is included. Writing the knowledge survey questions in Table 5.1 for the QL course took no more than a couple of hours.

The instructor should avoid asking tricky questions on a knowledge survey. For example, in a knowledge survey for the first course in a two-semester sequence, including a problem that appears to be solvable by methods of the first semester but actually requires a method not yet encountered could mislead students. They are likely to rate their confidence high, when they probably cannot do the problem. This brings us to consideration of the accuracy of knowledge surveys as a measure of student learning.

Accuracy of Knowledge Surveys

Nuhfer and Knipp (2003) introduced knowledge surveys as a means to assess changes in student learning and to improve course organization and planning. Their article contained a graph that suggested final exam scores in an astronomy class were similar to post-survey results. They observed that instructors can usually offer explanations for gaps in learning revealed by knowledge surveys, as the instructor (Dewar) was able to do for question #14 in the QL survey in Table 5.1 and Figure 5.1. Some (Wirth and Perkins 2005) embraced knowledge surveys as a course design and assessment tool, while others (Bowers et al. 2005) presented evidence that they did not reliably predict final grades or final exam performance. This disagreement brought up issues related to the reliability of exams (Nuhfer and Knipp 2005) and students' ability to assess their own learning accurately (Clauss and Geedey 2010; Ehrlinger et al. 2008; Kruger and Dunning 1999). Clauss and Geedey (2010) examined students' ability to self-assess in four upper division courses—one mathematics (Linear Algebra) and three biology (Aquatic Biology, Ecology, and Evolution). They found that students were more accurate self-assessors at low (knowledge) and high Bloom levels (analysis, synthesis, evaluation), and less accurate at intermediate levels (comprehension, application). The investigators commented that this pattern might be "caused by several factors, some rooted in learning and others in teaching practice, none of which were directly addressed in this study" (72). Other research (Bell and Volckmann 2011) suggested that confidence ratings on knowledge surveys are valid reflections of students' actual knowledge. However, students scoring high on the exams estimated their knowledge more accurately than the lower-scoring students, who were overconfident. Subsequently, Nuhfer et al. (2016) showed that widely used graphical conventions in the self-assessment literature introduce artifacts that invite misinterpretation and that not all numerical approaches used to investigate and describe the accuracy of self-assessment are equally valid.

Favazzo et al. (2014) examined whether knowledge surveys could assist with assessment of content-based student learning outcomes in a microbiology program. They found that requiring students to answer questions in addition to rating their confidence to answer them provided the most accurate results. This, of course, negates the time-saving benefit of the knowledge survey. They reported being "optimistic about the possible establishment of a knowledge survey platform to evaluate programmatic objectives" (257).

High ratings on knowledge surveys may not be foolproof evidence of student learning, but low ratings are strongly indicative of gaps in learning that could be investigated. Given their low burden in terms of time to construct and administer, their value in course planning, and their usefulness to students as a guide, knowledge surveys are worth serious consideration as one tool for triangulating evidence about changes in student learning.

> ### Pause: Knowledge Surveys—Questions to Consider
> - Is a knowledge survey a good match for the type of research question (*What works? What is? What could be?*) that I have?
> - Will I use the knowledge survey just for my study or will it serve other purposes such as course planning or a student guide?
> - What other evidence will I gather to triangulate my data?

REFERENCES

Adhikari, A., and D. Nolan. 2002. "But What Good Came of It at Last? How to Assess the Value of Undergraduate Research." *Notices of the AMS 49* (10): pp. 1252–7.

Bell, P., and D. Volckmann. 2011. "Knowledge Surveys in General Chemistry: Confidence, Overconfidence, and Performance." *Journal of Chemical Education 88* (11): pp. 1469–76.

Besterfield-Sacre, M., C. Atman, and L. Shuman. 1997. "Characteristics of Freshman Engineering Students: Models for Determining Student Attrition in Engineering." *Journal of Engineering Education 86* (2): pp. 139–49.

Bowers, N., M. Brandon, and C. Hill. 2005. "The Use of a Knowledge Survey as an Indicator of Student Learning in an Introductory Biology Course." *CBE Life Sciences Education 4* (4): pp. 311–22. doi: 10.1187/cbe.04-11-0056.

Bretz, S. 2014. "Designing Assessment Tools to Measure Students' Conceptual Knowledge of Chemistry." In *Tools of Chemistry Education Research*, edited by D. Bunce and R. Cole, pp. 156–68. Washington: American Chemical Society.

Clauss, J., and K. Geedey. 2010. "Knowledge Surveys: Students Ability to Self-Assess." *Journal of the Scholarship of Teaching and Learning 10* (2): pp. 14–24. Accessed August 31, 2017. http://josotl.indiana.edu/article/view/1742/1740.

Dewar, J. 2010. "Using a Knowledge Survey for Course and Program Level Assessment in Mathematics." Paper presented to the Joint Mathematics Meetings, San Francisco, CA, January 10–13. Accessed August 31, 2017. http://jointmathematicsmeetings.org/meetings/national/jmm/1056-f1-443.pdf.

Dewar, J., and C. Bennett, editors. 2015. *Doing the Scholarship of Teaching and Learning in Mathematics*. Washington: Mathematical Association of America.

Dewar, J., S. Larson, and T. Zachariah. 2011. "Group Projects and Civic Engagement in a Quantitative Literacy Course." *PRIMUS: Problems, Resources, and Issues in Mathematics Undergraduate Studies 21* (7): pp. 606–37.

Dommeyer, C.J., P. Baum, R.W. Hanna, and K.S. Chapman. 2004. "Gathering Faculty Teaching Evaluations by In-Class and Online Surveys: Their Effects on Response Rates and Evaluations." *Assessment and Evaluation in Higher Education 29* (5): pp. 611–23.

Draugalis, J., S. Coons, and C. Plaza. 2008. "Best Practices for Survey Research Reports: A Synopsis for Authors and Reviewers." *American Journal of Pharmacy Education 72* (1): pp. 11.

Ehrlinger, J., K. Johnson, M. Banner, D. Dunning, and J. Kruger. 2008. "Why the Unskilled Are Unaware: Further Exploration of (Absent) Self-Insight among the Incompetent." *Organizational Behavior and Human Decision Processes 105* (1): pp. 98–121.

Favazzo, L., J. Willford, and R. Watson. 2014. "Correlating Student Knowledge and Confidence using a Graded Knowledge Survey to Assess Student Learning in a General Microbiology Classroom." *Journal of Microbiology and Biology Education 15* (2): pp. 251–8. doi: 10.1128/jmbe.v15i2.693.

Fincham, J. 2008. "Response Rates and Responsiveness for Surveys, Standards, and the Journal." *American Journal of Pharmaceutical Education 72* (2), Article 43.

Fink, A. 2013. *How to Conduct Surveys: A Step-by-Step Guide*. Thousand Oaks: Sage Publications.

Finney, H. C. 1981. "Improving the Reliability of Retrospective Survey Measures: Results of a Longitudinal Field Survey." *Evaluation Review 5*: pp. 207–29.

Gieger, L., J. Nardo, K. Schmeichel, and L. Zinner. 2015. "A Quantitative and Qualitative Comparison of Homework Structures in a Multivariable Calculus Class." In *Doing the Scholarship of Teaching and Learning in Mathematics*, edited by J. Dewar and C. Bennett, pp. 67–76. Washington: Mathematical Association of America.

Hartley, J. 2014. "Some Thoughts on Likert-type Scales." *International Journal of Clinical and Health Psychology 14* (1): pp. 83–6.

Hartley, J., and L. Betts. 2010. "Four Layouts and a Finding: The Effects of Changes in the Order of the Verbal Labels and Numerical Values on Likert-type Scales." *International Journal of Social Research Methodology 13* (1): pp. 17–27. doi: 10.1080/13645570802648077.

Holcomb, J. 2015. "Presenting Evidence for the Field that Invented the Randomized Clinical Trial." In *Doing the Scholarship of Teaching and Learning in Mathematics*, edited by J. Dewar and C. Bennett, pp. 117–26. Washington: Mathematical Association of America.

Kaus, C. 2015. "Using SoTL to Assess the Outcomes of Teaching Statistics through Civic Engagement." In *Doing the Scholarship of Teaching and Learning in Mathematics*, edited by J. Dewar and C. Bennett, pp. 99–106. Washington: Mathematical Association of America.

Kruger, J., and D. Dunning. 1999. "Unskilled and Unaware of it: How Difficulties in Recognizing One's Own Incompetence Lead to Inflated Self-Assessment." *Journal of Personality and Social Psychology 77* (6): pp. 1121–34.

Mellor, B. 2015. "The Mathematics of Symmetry and Attitudes towards Mathematics." In *Doing the Scholarship of Teaching and Learning in Mathematics*, edited by J. Dewar and C. Bennett, pp. 157–70. Washington: Mathematical Association of America.

Moy, C., J. Locke, B. Coppola, and A. McNeil. 2010. "Improving Science Education and Understanding through Editing Wikipedia." *Journal of Chemical Education 87* (11): pp. 1159–62. doi: 10.1021/ed100367v.

Nuhfer, E., C. Cogan, S. Fleisher, E. Gaze, and K. Wirth. 2016. "Random Number Simulations Reveal How Random Noise Affects the Measurements and Graphical Portrayals of Self-Assessed Competency." *Numeracy 9* (1), Article 4. doi: 10.5038/1936-4660.9.1.4.

Nuhfer, E., and D. Knipp. 2003. "The Knowledge Survey: A Tool for all Reasons." In *To Improve the Academy*, Vol. 21, edited by C. Wehlburg and S. Chadwick-Blossey, pp. 59–78. San Francisco: Jossey Bass.

Nuhfer, E., and D. Knipp. 2005. "Re: The Use of a Knowledge Survey as an Indicator of Student Learning in an Introductory Biology Course." *CBE Life Sciences Education 5* (4): pp. 313–16. doi: 10.1187/cbe.06-05-0166.

Nulty, D. 2008. "The Adequacy of Response Rates to Online and Paper Surveys: What Can Be Done?" *Assessment & Evaluation in Higher Education 33* (3): pp. 301–14. doi: 10.1080/02602930701293231.

O'Hanlon, W., D. Barker, C. Langrall, J. Dossey, S. McCrone, and S. El-Zanati. 2015. "Mathematics Research Experiences for Preservice Teachers: Investigating the Impact on Their Beliefs." In *Doing the Scholarship of Teaching and Learning in Mathematics*, edited by J. Dewar and C. Bennett, pp. 171–82. Washington: Mathematical Association of America.

Poggio, J., M. Miller, and D. Glasnapp. 1987. "The Adequacy of Retrospective Judgments in Establishing Instructional Validity." *Educational and Psychological Measurement 47* (3): pp. 783–93.

Smith, J., E. Brown, D. Thoman, and E. Deemer. 2015. "Losing Its Expected Communal Value: How Stereotype Threat Undermines Women's Identity as Research Scientists." *Social Psychology of Education 18*: pp. 443–66. doi: 10.1007/s11218-015-9296-8.

Suskie, L. A. 1996. *Questionnaire Survey Research: What Works*, 2nd ed. Tallahassee: The Association for Institutional Research.

Szydlik, S. 2015. "Liberal Arts Mathematics Students' Beliefs about the Nature of Mathematics: A Case Study in Survey Research." In *Doing the Scholarship of Teaching and Learning in Mathematics*, edited by J. Dewar and C. Bennett, pp. 145–65. Washington: Mathematical Association of America.

Trytten, D., A. Lowe, and S. Walden. 2012. "'Asians are Good at Math. What an Awful Stereotype': The Model Minority Stereotype's Impact on Asian American Engineering Students." *Journal of Engineering Education 101* (3): pp. 439–68.

Wirth, K., and D. Perkins. 2005. "Knowledge Surveys: An Indispensable Course Design and Assessment Tool." Paper presented to the Conference on Innovations in the Scholarship of Teaching and Learning at Liberal Arts Colleges, St. Olaf, MN, April 1–3. Accessed August 31, 2017. https://www.macalester.edu/academics/geology/wirth/WirthPerkinsKS.pdf.

CHAPTER 6

Evidence

From Interviews, Focus Groups, and Think-Alouds

Introduction

When seeking to answer questions about science, technology, engineering, or mathematics (STEM) students' thinking, motivation, or attitudes, a scholarship of teaching and learning (SoTL) investigator should consider using the methods of gathering evidence discussed in this chapter—interviews, focus groups, and think-alouds—even if these methods are unfamiliar. That was certainly true for us when we began doing SoTL, but we have learned about them and put them to good use in our SoTL work. We begin with the most basic of these, the interview.

Interviews

Evidence gathered in **interviews** aligns well with questions that seek to know the "why" about something. Interviews are useful for discovering underlying reasons for factors such as participation, behavior, attitudes, persistence, and the like. They allow the researcher to probe a subject's motives, and because they are conducted one-on-one, they can be a better choice than a focus group if anything sensitive or personal is being discussed. The interviewer can use follow-up questions to seek further clarification or explanations and conversely, the subject can seek clarification of what is meant by a question. Neither of these is possible if a survey is used. In some studies, a survey of a larger population is followed up with interviews of a few participants.

Interviews can be **structured**, **semi-structured**, or **open**. In a structured interview, the exact same questions are asked in the exact same order of each subject. Some, or all, of the responses may be limited to specific choices or be open-ended. The semi-structured interview is the most commonly used format in SoTL research. In this type of interview the interviewer has a list of questions or target areas to be addressed. Generally, similar questions are asked of each subject, but additional follow-up questions can be asked as seem appropriate at the time. Some likely follow-up questions may be drafted in advance, while the interviewer may think of others during the interview, as a result of what the subject said. The subject can respond however they like to the questions. An unstructured interview has no set questions and proceeds much like a conversation about a topic. Creswell (2013) and Turner (2010) provide detailed information on preparing for and conducting an interview. The following sections summarize some of their points.

Preparing for an Interview

A good preparation for the interview is key to a successful outcome. The preparation process includes constructing the questions, choosing the site for the interview, practicing the interview protocol before finalizing it, recruiting and training the interviewers, selecting participants, and deciding on the method for recording the data.

Writing Good Questions

There are many different types of questions that can be asked. Patton (2002) offered a useful classification of six types of interview questions (Background/Demographic, Experience/Behavior, Opinion/Value, Feeling, Knowledge, Sensory) and Herrington and Dabuenmire (2014, 42) gave examples from chemistry education research for each of Patton's question types. There is no need to use every question type; rather the goal is to write prompts that will help answer the research question. Before writing any questions, it is a good idea to outline the specific topics that the interview is intended to explore. Ways of testing whether the interview questions directly relate to the research question include: imagining the possible responses they might elicit, asking a colleague to read and comment on the questions, or conducting pilot interviews with a few people to see if the responses are useful.

Many of the guidelines given for writing survey questions (Chapter 5, Essential Steps in Designing a Survey) apply to interview questions. We reiterate a few of them here.

- Avoid leading questions, such as "What did you like about (teaching method)?" A neutral version of the same question would be "Tell me what you thought about (teaching method)?"
- Questions should be clear and not address more than one thing at a time. For example, "How would you describe your motivation and effort?" is really two different questions, one about motivation and one about effort.
- Be wary of "why?" questions. Asking a student why they gave a particular answer may make them feel there was something wrong with their answer. Generally, why questions can be rephrased to something less intimidating such as "Can you tell me more about your thought process in answering that question?" (Herrington and Dabuenmire 2014, 43).

Conducting an Interview

How the interview is conducted can also make a difference in the quality of the data obtained. The interview should begin with a description of its purpose, who will have access to the answers, and how the answers will be analyzed. Ask for permission to record the interview or to take notes. Generally, it is better to record the session, as it is impossible to take notes of everything that is said. This type of human subjects research will usually require the researcher to obtain informed consent as described in Chapter 3 (Human Subjects Considerations). If so, be prepared to request that participants sign the consent form approved by the local Institutional Review Board and to explain that they can stop participating at any point. Seidman (2013, 64–77) discussed in detail how to adapt the requirements for informed consent for in-depth interviewing research.

Indicate how long the interview usually takes and ask if the person has any questions before starting the interview. The interviewer needs to put the subjects at ease so that they feel they can respond truthfully without being "judged." Because of the inherent power differential between students and their instructor, an interview about a specific course's content or teaching methods is best conducted by someone other than the instructor of the course. Students should be told when their instructor or others will have access to the interview data (typically after grades are submitted if the interview is course-related) and whether identifying characteristics will be retained or removed.

There are several steps the interviewer can take to encourage more detailed responses. The simplest is to allow several seconds of silence to pass before moving on to another question. A simple nod of the head or an occasional "uh-huh" will also encourage the subject. Another technique is to repeat what was said, "What I heard you say was that you…" and then pause.

Follow up questions can also elicit additional details or clarification. Examples are: "Can you tell me more about that?" "Can you give an example?" "Can you describe…more fully?" It is important for the interviewer to remain neutral and not to appear surprised, pleased, or disturbed by a response. Note taking, if done at all, should be as inconspicuous as possible. It is easy to imagine that a subject could be distracted or upset if the interviewer suddenly started scribbling furiously after some response.

Signaling transitions with statements such as, "We have been talking about (one particular topic), now let's move on to (another topic)," can help keep a subject on track and focused. If subjects stray to other topics, the interviewer can bring them back with a similar statement. Closing the interview with a general invitation to say more, "Is there anything else you would like to say?" allows the subject to add or ask anything.

Depending on the topic under consideration, at the end, the interviewer can further reassure the student with a comment like this: "I'll take care to write this up in a way that will not identify you. Have you told me anything that I need to take special care with, that I should withhold, or that you would like to see how I have worded it before I use it?"

Obtaining a Transcript of the Session

In order for the recording of the session to be analyzed, it needs to be transcribed, which takes either time or money. Because people talk much faster than a person can type, a single hour of recording can take three or four hours to transcribe. Transcription services will do this task for a fee. They may charge by the length of the recording, the time required to transcribe it, or the length of the transcript produced. Expect the cost to run on the order of $1.50–$2.50 per minute of recording, depending on the quality of the audio, the number of voices, how technical the material is, and the degree of exactness desired (that is, are "ums," "ers," "pauses," and repeated words to be included?).

However, if the researcher has the time and inclination to do the transcription, there is much to be gained beyond saving money. The transcription process immerses the transcriber in the data and becoming familiar with the data is key to its subsequent analysis. Recent advances in technology can assist with transcription. For example, improvements in **voice recognition software** (VRS) allow the researcher to listen to a segment of a recording and then repeat the segment into a VRS program to produce a transcript. A foot-pedal or other device to toggle easily between play and pause helps whether one is using VRS or typing directly. Breaking up transcribing sessions into smaller chunks of time, an hour or less in one sitting, will improve efficiency overall.

However the transcription is done, when writing up the work for publication it is important to be transparent about the process, the decisions that were made, and the meaning of any notation being used. An excerpt from Trytten et al. (2012) provides a good illustration of this:

> We removed conversational fillers, such as "ummm," "like," and "you know." Words not said by the participant that were inserted or substituted by the authors to clarify context are given in square brackets ([ ]), a practice also used in a literature quote above to remove a racially insensitive word. When words or sentences are deleted, ellipses (…) are used. Words that are not clear upon transcription and verification are put in parentheses. Words in angle bars (< >) are categorizations substituted to avoid identifying participants. A (p) indicates that the participant paused at that point. Statements preceded with a "P:" were said by the participant. Statements preceded with an "I:" were said by the interviewer (449).

These suggestions for transcription also apply to audio recordings gathered in focus groups or think-alouds.

Once the transcript is obtained, the researcher studies it for insights or examples, using disciplinary knowledge, applicable theoretical frameworks, or existing taxonomies. Coding transcript data for themes is a standard approach to analyzing this type of data. This content analysis process is described in detail in Chapter 7 (Coding Data).

Examples of STEM Education Studies that Used Interviews

In a qualitative study, Gibau (2015) used transcripts from 24 individual interviews and one focus group to examine the perspectives and experiences of underrepresented minority (URM) students engaged in National Institutes of

Health-funded intervention programs and to assess program effectiveness from these students' accounts. Questions of interest in this study included: What were the experiences of the URM students? Which experiences were most critical to their persistence in their programs? And how can these experiences potentially inform and enhance intervention strategies? As Gibau noted, qualitative studies like this one "are particularly valuable, since they offer a more nuanced viewpoint on assessing the efficacy of such interventions and serve to complement existing quantitative research" and they "provide a critical contribution to ongoing conversations regarding underrepresentation in the sciences."

Generally, interviews seek information about attitudes, beliefs, opinions, or experiences, but interview questions can also explore technical or disciplinary knowledge. Herman et al. (2012) used open-ended interview questions to reveal electrical engineering, computer engineering, and computer science students' misconceptions about state in sequential circuits. Because the goal was to examine students' misconceptions, the researcher chose students who had completed a digital logic course with grades of B or C, but not A. Students were asked a series of definition questions, relationship questions, and design problems. The misconceptions that the researchers identified suggested implications for instruction and future research and provided a basis for constructing a Digital Logic Concept Inventory. When students are instructed to verbalize their thinking as they answer questions, the interview becomes more like a think-aloud, which we discuss later in this chapter (Think-Alouds). Weinrich and Talanquer (2015, 2016) employed this type of interview to investigate students' modes of conceptualizing and reasoning when thinking about chemical reactions used to make a desired product.

Pause: Interviews—Questions to Consider

- Is an interview a good match for the type of research question (*What works? What is? What could be?*) that I have?
- Will I conduct the interview, or should I recruit someone else to conduct it?
- Have I thought about where to conduct the interview, who the subjects would be, and how to recruit them?
- Have I followed the recommendations for preparing for the interview, including having a practice or pilot interview?
- Is the interviewer familiar with techniques for encouraging more detailed responses?
- How do I plan to get a transcript of the interview?

Focus Groups

A **focus group** is essentially a group interview structured so as to obtain the spread of opinions on an issue from a group of individuals. Unlike a one-on-one interview, a focus group allows participants to hear and interact with one another. It seeks a range of opinions, views, or perceptions, not a consensus. The number of participants can vary from as few as six to a dozen or more. One focus group takes far less time (typically 60 to 90 minutes) than conducting one-on-one interviews with each of the subjects. However, a personal interview would be the better choice if the topic to be explored is sensitive or if the researcher desires to explore the topic deeply with each individual subject. A focus group can provide a clearer picture of what lies underneath student opinions than an open-ended survey question can.

Model for Conducting a Focus Group

Krueger and Casey (2014) wrote a very practical and readable guide to focus groups for novice researchers. However, in this section we follow the model provided by Millis (2004), as it is a very efficient process for conducting a focus group and analyzing and reporting the findings. The process and the materials needed are as follows:

Step 1. Preparation.

- Find an impartial facilitator, someone not involved with the study, to conduct the focus group. The person should have some experience with conducting interviews, if not focus groups. Be sure that the person understands the procedures to be followed.
- Optional: Get an assistant facilitator to take notes or arrange to have the focus group recorded.
- Develop the questions to be asked based on careful consideration of the information desired.
- Recruit participants, typically students, whose perspectives are sought. Students can be offered pizza or snacks (before or after, not during) to participate. Let participants know that the focus group will start exactly on time, and there will be no late admissions.
- Optional: Prepare a short survey on paper for students who arrive early.
- Prepare a script for the facilitator, the presentation slides (containing the directions for the procedure outlined in Step 2), the roundtable

ranking form (one for each smaller team as noted in Step 2), the initial survey (if one is to be given to the early arrivals), and index cards for each participant.
- Have the seating arranged in a circle or horseshoe-shape.
- Start on time, close the door, and put a notice on it that late arrivals cannot enter. Everyone has to be present at the beginning.
- Turn on and check the recording device, if using one.

Step 2. The Focus Group Procedure.

- Welcome and thank everyone, make introductions, and explain the purpose of the focus group—to gain a range of opinions about the topic.
- Explain the ground rules. The focus group is:
  - *Anonymous*: Participants' comments will not be attributed to them individually. They will use an assigned number when making comments and referring to other participants.
  - *Candid*: Honest, constructive information is sought. All comments, positive and negative, are useful in gaining insight about the topic under discussion.
  - *Confidential*: Only appropriate parties get the results. Participants should not discuss comments afterwards.
  - *Productive*: Indicate to the participants how the information will be used.
- Index card rating activity
  - Each participant takes an index card.
  - Participants count off #1, #2, etc., to get their numbers.
  - Each participant puts his or her number in the upper right-hand corner of the index card and announces the number each time before speaking.
  - On the index card, each person writes a word or phrase to describe his or her impression of the topic being explored by the focus group. Below that word or phrase, participants are asked to write a number from 1 to 5 that describes the level of their satisfaction related to the topic, where 1 indicates low satisfaction and 5 indicates high.
  - Once the writing and rating are completed, participants are directed to report to the whole group as follows (while the assistant takes notes):
    - "Would a volunteer please read your phrase and rating?" Remind the speaker to state his or her number before speaking.

- "Please tell why you wrote that phrase and rated the way you did."
  - Prompt a next volunteer with, "Can we hear from someone else?" (Remind the speaker to state their number before speaking.)
  - Continue until all speak or signal an anticipated end with, "Let's hear from two more."
- Collect the cards.
- Roundtable generating and ranking activity
  - Going around the circle, divide the group into teams of three or four adjacent participants.
  - Provide each team with a sheet of paper or pre-prepared roundtable ranking form. (Appendix III contains a sample roundtable ranking form.) Give directions to pass the sheet rapidly from person to person with each person writing down one strength or positive aspect of the intervention or topic under study, saying it aloud while writing it down. They should put their participant numbers next to what they write. An individual can say "Pass" without writing.
  - Working as a team, members then rank-order the top three strengths they identified, with the most important strength ranked first.
  - Using the other side of the sheet of paper or form, the generating and ranking process is repeated for drawbacks or things that could be improved. The most important thing to improve is ranked first.
  - The papers are collected.
- Open-ended question(s)
  - The remaining time can be devoted to full group consideration of several open-ended questions. They should be presented in a logical order.
  - As before, participants state their numbers before speaking.
  - Responses can be structured as a round-robin, that is, starting at some point in the circle, and going around from person to person to get a response. Anyone can pass. This approach ensures that everyone has a chance to speak, and avoids having to solicit volunteers.
- Wrap-up
  - As a closing activity, group members can be invited to say one thing they heard that was really important.
  - Thank the participants and remind them of the confidentiality of what they heard and how the information will be used.

Step 3. Analyze the Results.

- The index card activity produces a set of numbers (ranging from one to five) to average and a collection of short phrases to code or categorize as described in the Coding Data (Content Analysis) section in Chapter 7.
- The teams in the roundtable ranking activity have identified and rank-ordered strengths and weaknesses, which can be coded for common themes (see Chapter 7, Coding Data).
- The recordings or notes taken can be transcribed, if a full record of the discussion is desired, and the transcripts can be coded.

Millis (2004) wrote that a typical transcript entry "might look like this: 'Student Seven: Unlike Student Two, I felt the evaluation methods were extremely fair because the essays gave us a chance to demonstrate what we actually knew. I found myself integrating ideas in ways I hadn't thought of before'" (130). She advised paying for a transcription service to produce the transcripts. The methods for transcribing interview recordings that we discussed earlier in this chapter (see previous section: Interviews, Obtaining a Transcript of the Session) apply to focus group recordings as well. The appendices in her paper show how to represent the data from the index card and roundtable activities as histograms and color-coded tables.

Examples of STEM Education Studies that Used Focus Groups

Dewar and her colleagues (Dewar et al. 2011) found Millis's model easy to implement when they conducted a focus group for their study of incorporating group projects on local community issues into a quantitative literacy (QL) course. Sample materials from their QL study focus group, including the script, the presentation slides, roundtable ranking form, and initial survey questions, are found in Appendix III.

Focus groups are often used to gain qualitative data that builds on data obtained from a survey. For example, Mostyn et al. (2013) learned from a survey of nursing students who had access to podcasts of a biological science course that 71 percent of respondents had accessed at least one podcast; and further, 83 percent of students who reported accessing podcasts agreed that they were useful as learning tools. Focus groups were then used to explore student perceptions of the usefulness of the podcasts for their learning. The researchers report running the focus groups in a manner very similar to Millis' model. In particular, they were conducted

> over a lunchtime period with refreshments being provided...by a researcher who was independent from the students' teaching team and a second independent researcher...acted as an observer and took notes with respect to body language and group interactions. The focus groups lasted for 90 minutes each. Discussions were digitally recorded...and the recordings were transcribed verbatim. (Mostyn et al. 2013, 3)

Here are several more examples of STEM education studies that employed focus groups (though not necessarily Millis's model) to gather data. To examine mechanical engineering students' perceptions of how effectively an instructor can express ideas using a tablet PC (Walker et al. 2008), a focus group was conducted following an in-class survey. Results of this study suggested that students were more likely to pay attention during the lecture and recognize the more salient points of the presentation when a tablet PC was used.

Kearney et al. (2015) investigated group/team development in computer engineering courses from the perspective of organization behavior theory. They used a combination of linguistic analysis on student reflection essays and focus groups as a method to obtain information about how student groups and teams functioned.

Focus groups proved useful in program assessment efforts conducted by the Chemistry Department at the University of Scranton in collaboration with the Assessment and Institutional Research Office and the Department of Counseling and Human Services (Dreisbach et al. 1998). As Dreisbach et al. noted, the use of direct conversation, whether in a focus group or an interview, yields less ambiguous results compared with other methods because responses can be clarified with careful follow up questions.

Pause: Focus Groups—Questions to Consider

- Is a focus group a good match for the type of research question (*What works? What is? What could be?*) that I have?
- Have I followed the recommendations for preparing for the focus group, including finding an impartial facilitator?
- Have I thought about where the focus group will be conducted, who the participants would be, and how to recruit them?
- If the focus group will be audio-recorded, how do I plan to get a transcript of the recording?

Think-Alouds

A **think-aloud** is an interview-like method developed by social science researchers to study mental processes. Subjects are given a task and instructed to talk out loud as they perform the task. They are audio- or video-recorded. After completing the task, subjects can be asked to reflect back on and report all they remember about their thinking. The validity of these verbal reports as evidence of the subjects' thinking rests on their ability to verbalize the information that they are paying attention to as they perform the task. In other words, at the core of this method is the hypothesis that "the information that is heeded during the performance of a task is the information that is reportable; and the information that is reported is information that is heeded" (Ericsson and Simon 1993, 167). Similar to the situation with interviews, because of the position of power instructors hold over students, they should not conduct think-alouds with current students.

We begin with some examples of think-aloud studies conducted in engineering, physics, and chemistry education and then describe in some detail how two of the authors (Bennett and Dewar) used the method to study student understanding of mathematical proof. We also offer specific advice and models for procedures to follow to elicit accurate reporting.

An Example from Engineering Education

Litzinger et al. (2010) recorded students as they talked out-loud while solving problems in statics. The research questions they wanted to investigate were:

1. What elements of knowledge do statics students have the greatest difficulty applying during problem solving?
2. Are there differences in the elements of knowledge that are accurately applied by strong and weak statics students?
3. Are there differences in the cognitive and metacognitive strategies used by strong and weak statics students during analysis? (337).

Because these are *What is?* questions that seek to understand what it is that students do when they solve statics problems, using think-aloud sessions during which students solved typical textbook problems was an appropriate choice. They selected the work of twelve (out of 39) students for detailed analysis. Of the twelve, six had been identified as weak and six as strong problem solvers, based on scores on a course exam and the think-aloud problems. The think-aloud data from the two sets of students were analyzed to identify common technical errors and any major differences in the problem-solving processes. The think-alouds revealed that the weak, and most of the strong, problem solvers relied heavily on memory to decide what reactions were present at a given connection, and few of the students could reason physically about what reactions should be present. The researchers also found substantial differences between weak and strong students in the use of self-explanation (that is, explaining how a specific concept applies to a portion of the problem, such as identifying a reaction force due to the restriction of movement or recognizing the reaction forces associated with a particular type of connection).

In order to conduct their study, Litzinger and his colleagues constructed a model of the problem-solving process, which was based on the problem-solving literature (McCracken and Newstetter 2001; Van Meter et al. 2006), to analyze the approaches used by the students. Their Integrated Problem-Solving Model is presented as a table listing three phases of problem solving—recognition, framing, and synthesis—and for each phase identifies the problem-solving processes, prior knowledge, and symbol systems used. This model provided the basis for the rubric used to evaluate and code participants' responses. In Chapter 7 (Rubrics) we discuss the development and use of rubrics in some detail.

Litzinger and his colleagues noted that they followed the recommendations from Ericsson and Simon (1993) and Pressley and Afflerbach (1995) for conducting a think-aloud in order to secure the most reliable and valid reports possible: "First, reports were concurrent with learner processing, which means that the students verbalize what they are thinking while solving the problems. Second, participants were given only general prompts (e.g., 'tell what you are thinking,' 'tell what you are doing') and instructions did not refer to any problem specifics. Third, participants completed a practice problem while thinking aloud to begin the session" (Litzinger et al. 2010, 340).

An Example from Physics Education

Disappointing results in a *What works?* experiment prompted a *What is?* study (Bujak et al. 2011) to determine the issues underlying the poor outcome. A new curriculum designed to improve student learning by organizing concepts around fundamental principles had failed to produce the desired results: on force concept questions, the students taught in traditional classes continued to outperform those in the new curriculum. To understand what principles and processes students were using when solving the force concept inventory problems a think-aloud method was used. They found an almost complete absence of the word "momentum" or the phrase "change in momentum" despite momentum being one of three core principles around which the curriculum was designed. Apparently, students had failed to grasp the new framework and had not gained an understanding of how physical systems behave.

Using Think-Alouds in Chemistry Education Research

An overview of how a think-aloud might be used to investigate student thinking in introductory chemistry appeared in the *Journal of Chemical Education* (Bowen 1994). Bowen noted that think-aloud methods "have a long history of use in chemistry education" (190). Using a hypothetical study, he illustrated how to develop a research question and strategy, prepare for and conduct think-alouds or interviews, analyze the data, and present the results. His discussion of how to present qualitative data that results from such a study is worth attention.

Examples from Biology Education

In the context of exploring how students understood the central dogma of molecular biology when presented in the canonical representation, Wright et al. (2014) employed think-alouds. Participating students were shown the representation and asked to explain their interpretation of the diagram as a whole and of any individual processes that were included. The think-aloud interviews were used to complement written explanations of the same diagram created by a larger number of students.

Southard et al. (2016) used think-alouds as part of a project focused on characterizing student understanding of key principles in molecular and cellular biology. Students were asked to create concepts maps for important ideas in molecular and cellular biology and to think aloud as they created the concept map. They were also asked to provide verbal explanations of connections between concepts in their finished maps. The verbal comments by students were used to characterize the connections they made between different concepts.

An Example from Mathematics Education

The think-aloud method proved effective when Bennett and Dewar (2007, 2013) investigated the evolution of students' understanding of proof as they progressed through the Loyola Marymount University mathematics curriculum. As 2003–2004 Carnegie scholars and experienced college mathematics instructors, they wanted to investigate the evolution of students' understanding of mathematical proof as they progressed through the mathematics curriculum and to identify the learning experiences that promoted growth in their understanding. Their Carnegie mentors suggested using a think-aloud methodology and referred them to Ericsson and Simon (1993) for details. Eventually, think-alouds were conducted with 12 students in which they first investigated and then attempted to prove a simple conjecture from number theory. Dewar conducted the think-alouds during a sabbatical year, so none of the subjects were her current students.

Four courses in the curriculum that involved proof were identified and at least two students who were at points in the curriculum just before and after these courses were recruited to participate. A faculty member was selected to serve as an expert subject for the think-aloud. The subjects were given pencil and paper and instructions to think-aloud as they investigated the following mathematical situation:

> Please examine the statements:
> For any two consecutive positive integers, the difference of their squares
>   (a) is an odd number, and
>   (b) equals the sum of the two consecutive positive integers.
> What can you tell me about these statements?

Even if they made a general argument as part of their investigation, subjects were asked to try to write down proofs for their conjectures. Finally, students were given tasks to evaluate several sample proofs for correctness, to check whether they would apply the proven result, and to see if they thought a counterexample was possible after having a valid proof. Throughout, students were reminded to keep talking aloud.

The resulting think-aloud transcripts contained significantly richer data than could have been obtained from work written by the students while they performed the task. In particular, they contained evidence that there were affective influences on student performance, that students might be able to employ sophisticated mathematical thinking but be unable to communicate it in writing, and that having additional knowledge might result in poorer performance on the task. From this rich data the researchers were able to develop a detailed taxonomy of mathematical knowledge and expertise (Bennett and Dewar 2007, 2013) that drew on a typology of scientific knowledge (Shavelson and Huang 2003) and Alexander's (2003) model of domain learning. The end result went far beyond answering the question about the evolution of students' understanding of proof.

This particular SoTL project exemplifies the interdisciplinary nature of SoTL work in several ways. The findings resulted from employing disciplinary knowledge in combination with social science methodology, a typology from science education, and a model from pre-collegiate education adapted to college-level mathematics. Subsequently, the taxonomy served as a model for describing an "arc" of progression through an undergraduate computer science curriculum based on open source principles, values, ethics, and tools (Dionisio et al. 2007).

The project also provides an example of theory-building occurring as an unintended outcome of a SoTL project. This contrasts with the work of Litzinger et al. (2010) discussed in this chapter (Think Alouds, An Example from Engineering Education) that began by consulting the literature to develop a model for the assessment of student work in a think-aloud on problem-solving in statics.

Advice for Conducting a Think-Aloud

Whether conducting think-alouds with a single subject or with multiple subjects, a detailed plan or script, sometimes called a **protocol**, should be prepared for use. The think-aloud script should begin with specific instructions for greeting the subject, explain in general terms the purpose for the think-aloud, describe thinking aloud, and supply practice exercises so that the subject can experience thinking aloud. Subjects should practice reporting their thinking both concurrently and retrospectively (Ericsson and Simon 1993). The **concurrent report** consists of what they say as they think-aloud while they perform the task. The **retrospective report** is obtained after they have completed the task by asking them to report anything else they can remember about their thinking while they did the task.

Sample Warm-up Script for a Think-Aloud

We provide a sample warm-up script, derived from one found in Ericsson and Simon (1993, 377–8) and similar to the one used in Bennett and Dewar (2007, 2013). It contains practice exercises and requests both concurrent and retrospective reports. The words the person conducting the think-aloud speaks to the subject appear in italics.

Thanks for agreeing to be interviewed. The purpose of this interview is to take a close look at mathematics students' thinking processes. Your responses are completely voluntary and do not affect any coursework or grade. They will be kept confidential. If any portion of your response is made public, a pseudonym will be used so that you cannot be identified. You will be asked to do a "think-aloud" while you are working on a problem or question. That is, I want you to tell me everything you are thinking from the time you first see the question until you give an answer. I would like you to talk aloud constantly from the time I present each problem until you have given your final answer. I don't want you to try to plan out what you say or try to explain to me what you are saying. It is not necessary to use complete sentences. Just act as if you are alone in the room speaking to yourself.

At this point the subject can be given the informed consent form and asked to sign it, by saying something like: *in order to include your responses in the study, your written permission is needed on this consent form.* The person conducting the interview can then summarize any additional details from the consent form and assure the subject that he or she may stop participating at any time during the think-aloud.

Thank you, first we will practice having you think aloud. Please face the desk. I'll sit behind you so that it seems more like you are alone and thinking aloud. It is most important that you keep talking. I will occasionally remind you to keep talking if you are silent for too long. Do you understand what I want you to do?

Good. Here is the first practice question: how many windows are in the house or apartment where you live?

After the subject completes the practice think-aloud task, Ericsson and Simon (1993) suggested that the researcher request a retrospective report, which can be done in the following way:

Good, now I want to see how much you can remember about what you were thinking from the time you heard the question until you gave the answer. I am interested in what you actually can remember, rather than what you think you might have thought. If possible, I would like to tell me about your memories in the sequence in which they occurred while working on the question. Please tell me if you are uncertain about any of your memories. I don't want you to work on solving the problem again, just report all you can remember thinking about when answering the question.

Now tell me all that you can remember about your thinking.

Good, now I will give you two more practice problems before we proceed. I want you to do the same thing for each of these. I want you to think aloud as before as you think about the question. After you have answered it, I will ask you to report all that you can remember about your thinking. Any questions?

Additional practice exercises suggested by Ericsson and Simon include:

- Name 20 animals (subjects were told the researcher would keep track of the number).
- Solve the anagram ALIOSC.

Because our task was mathematical, we wrote some mathematically oriented practice problems:

- If the sum of two whole numbers is odd, what can you conclude about their product, and why?
- If a right triangle has hypotenuse 25 and leg of 24, find the other leg.

The subject was told that pencil and paper could be used on the mathematical tasks. We alternated between the two types practice exercises until the subject indicated that was enough practice; one of each sufficed for most participants.

As the sample script indicates, the subject should be positioned facing away from the researcher. Sitting face-to-face makes it seem more like an interview to the subject. If appropriate, the subject can have pencil and paper or other tools available for use and be seated at a desk. The choice of what, if any, tools to provide should be made carefully, as the availability of a particular tool such as a graphing calculator, or lack thereof, might influence how the subject approaches a task.

If the subject remains silent for too long, more than a few seconds, gently remind the person to keep talking out loud by saying something neutral like, "remember to keep talking out loud" or "please tell me what you are thinking." This prompt is key to eliciting as complete a record of the subject's thinking as possible.

The protocol script should cover the entire think-aloud, not just the initial practice task. A well-prepared script has the tasks aligned with the questions being investigated. In the case of multiple subjects, having the script helps the researcher ask each subject the same questions, in the same order.

More Considerations when Conducting a Think-Aloud

Ericsson and Simon (1993) noted that sometimes subjects give descriptions of the problem-solving activity rather than thinking aloud. One indicator that subjects are describing rather than reporting is using the progressive tense, saying things like, "right now I am looking at equations 2 and 3 and thinking of applying..." (244). This is not a report of the information the subject is paying attention to during the task; it is a response to the internal query, "what am I doing?" Speaking at a slower rate is another indicator of giving a description. Ericsson and Simon considered these sorts of reports as less reliable.

Pressley and Afflerbach (1995) wrote that Ericsson and Simon (1993) strongly advised against asking subjects to explain why they were doing what they were doing, as this would invite theorizing about their cognitive processes (131). However, some STEM education studies that employed think-alouds did report doing exactly that (e.g., Bujak et al. 2011). Others suggested giving more specific prompts or asking for clarification with comments like "I'm not sure what you mean by that. Would you please explain that to me?" (Bowen 1994, 188). Austin and Delaney (1998) reviewed methods for eliciting reliable and valid verbal reports and supported the recommendation of telling subjects to think aloud but not to explain their behavior. Thus, we suggest that any prompts asking for explanations be reserved for interviews.

A researcher can gain more confidence in conducting think-alouds by experiencing one, that is, talking out loud while doing the practice tasks and the task the subjects will be asked to do. Conducting a practice or pilot think-aloud is a good idea. Also, a practical guide by van Someren et al. (1994) based on material from a University of Amsterdam course on the think-aloud method may be helpful.

As a practical matter, be sure that the audio or video recording device used has sufficient battery power for the entire session. Once the session is recorded, the next step is to download the recording as a digital file to a computer. It is important to choose a recording device that is reliable and compatible with the intended method of storage, transcription, and analysis.

Conducting a think-aloud is a time-intensive undertaking. Some researchers have gathered evidence of student thinking by asking them to write down their thinking as they solved a problem rather than conducting think-alouds with students one at a time. The study by Prevost and Lemons (2016) to understand better the procedures students used in answering multiple-choice problems in an introductory biology course provides an example. Each exam in the course included four problems where students were asked to write down what they thought as they worked to answer the problems. Prior to taking the exams, the instructor had modeled in class the process of writing down the thinking involved in answering a multiple-choice question. Students then practiced it on homework problems, received feedback, and were shown examples of good documentation and poor documentation of one's thinking. The written responses from the exams were analyzed by a mixture of qualitative and quantitative methods.

This method for gathering evidence of student thinking minimizes the time required to collect the data and avoids the need to transcribe the recording. However, the underlying assumption of Ericsson and Simon (1993) that think-aloud subjects are able to verbalize the information that they are paying attention to as they perform the task may not apply when subjects are directed to write their thoughts down. In addition, there is the possibility that the process of writing may introduce other interfering factors.

> **Pause: Think-Alouds—Questions to Consider**
>
> - Is a think-aloud a good match for the type of research question (*What works? What is? What could be?*) that I have?
> - Have I selected an appropriate task for the think-aloud and prepared a protocol?
> - Have I thought about where the think-aloud will be conducted and with whom?
> - Will I conduct the think-aloud myself, or should I recruit someone else to do it because I am currently teaching (some of) the subjects?
> - Will there be a practice or pilot think-aloud?
> - How do I plan to get a transcript of the recording of the think-aloud?

Student Voices in the Scholarship of Teaching and Learning

Before closing this chapter, which focused on three methods of gathering evidence that would help us understand what students were doing, thinking, or feeling, we pause to consider the key role that **student voices** play in SoTL. Because SoTL focuses on student learning, this form of scholarship necessarily engages us in work that involves students. Without evidence gathered from students how could we draw conclusions? But some SoTL researchers are going farther than just listening to students by taking on students in substantial roles as co-investigators (Cliffe et al. 2017).

An Example from Chemistry Education

One of the themed groups participating in Carnegie's Institutional Leadership and Affiliates Program (See Chapter 1, Forging Disciplinary Connections for

SoTL) was organized around student voices in the scholarship of teaching and learning and a commitment to involving students in collaborative efforts to improve teaching and learning. An edited volume on engaging student voices in the scholarship of teaching and learning (Werder and Otis 2010) emerged from this group. One of its chapters (Drummond and Owens 2010) was based on an investigation of student understanding of the chemical concept of charge distribution. A small group of first-year chemistry students were videotaped as they worked together to understand this concept. The instructors identified a particularly puzzling portion of the recording and converted it into still images with captions taken from the transcribed dialogue. This form of documentation allowed the students, their instructor, and other faculty to access the voices of the students in the group and enabled them to co-construct the meaning of the students' experience. Drummond and Owens testified to the value of this investigation for everyone involved in or reflecting on the study—students in the video-capture, students in the class but not in the capture, the instructor, and other faculty.

Students as Pedagogical Researchers

While undergraduate research has been a growing enterprise for a number of years, particularly in STEM fields, Partridge and Sandover (2010) noted that there are far fewer examples of undergraduates doing pedagogical research. They described a program at the University of Western Australia, a research-intensive institution, designed to support undergraduate research on topics related to teaching and learning at the university level. They reported several unexpected benefits of undergraduates undertaking pedagogical research that "may not have occurred in a discipline-based undergraduate research exercise." Among them were "cross-institutional involvement [that] allowed students an opportunity to interact with and learn from students in different disciplines" and "students reported improved relationships not only with the staff supporting them in their research but also with their teachers as a whole." Partridge and Sandover (2010) concluded that undergraduates are indeed capable of participating in pedagogical research as investigators. They also cautioned the need to avoid "tokenism:"

> [E]xpectations of quality of undergraduate research should be high, of a standard sufficient for publication and/or conference presentation. The focus of the research should be meaningful and relevant to both the student and a wider

audience. A rigorous and structured preparation in basic research methods and ongoing supervision are essential elements to ensure successful outcomes for the individual students and the institution.

> **Pause: Gathering Evidence—Questions to Consider**
>
> - If I plan to examine the student experience in my study, what evidence will I gather?
> - Would evidence gathered from interviews, focus groups, or think-alouds align well with my question?
> - Am I collecting more than one type of evidence in an attempt to triangulate my data?
> - To what degree do I want student voices reflected in my study? Would it be appropriate to take on one or more students as co-investigators?
> - What campus resources or collaborators might I seek out?

REFERENCES

Alexander, P. 2003. "The Development of Expertise: The Journey from Acclimation to Proficiency." *Educational Researcher 32* (8): pp. 10–14.

Austin, J., and P. Delaney. 1998. "Protocol Analysis as a Tool for Behavior Analysis." *The Analysis of Verbal Behavior 15* (1): pp. 41–56.

Bennett, C., and J. Dewar. 2007. "Developing and Applying a Taxonomy for Mathematical Knowledge-Expertise." *Electronic Proceedings for the Tenth Special Interest Group of the Mathematical Association of America on Research in Undergraduate Mathematics Education Conference on Research in Undergraduate Mathematics Education,* San Diego CA. Accessed August 30, 2017. http://sigmaa.maa.org/rume/crume2007/eproc.html.

Bennett, C., and J. Dewar. 2013. "SoTL and Interdisciplinary Encounters in the Study of Students' Understanding of Mathematical Proof." In *The Scholarship of Teaching and Learning In and Across the Disciplines,* edited by K. McKinney, pp. 54–73. Bloomington: Indiana University Press.

Bowen, C. 1994. "Think-Aloud Methods in Chemistry Education: Understanding Student Thinking." *Journal of Chemical Education 71* (3): pp. 184–90. doi: 10.1021/ed071p184.

Bujak, K., R. Catrambone, M. Caballero, M. Marr, M. Schatz, and M. Kohlmyer. 2011. "Comparing the Matter and Interactions Curriculum with a Traditional Physics Curriculum: A Think Aloud Study." *Proceedings of the 2011 AERA Annual Meeting,* New Orleans, LA, 2011. Accessed August 30, 2017. https://arxiv.org/ftp/arxiv/papers/1011/1011.5449.pdf.

Cliffe, A., A. Cook-Sather, M. Healey, R. Healey, E. Marquis, K. Matthews, L. Mercer-Mapstone, A. Ntem, V. Puri, and C. Woolmer. 2017. "Launching a Journal About and

Through Students as Partners." *International Journal for Students as Partners 1* (1). doi:10.15173/ijsap.v1i1.3194.

Creswell, J. 2013. *Qualitative Inquiry and Research Design: Choosing Among Five Approaches.* Thousand Oaks: Sage.

Dewar, J., S. Larson, and T. Zachariah. 2011. "Group Projects and Civic Engagement in a Quantitative Literacy Course." *PRIMUS: Problems, Resources, and Issues in Mathematics Undergraduate Studies 21* (7): pp. 606–37.

Dionisio, J., L. Dickson, S. August, P. Dorin, and R. Toal. 2007. "An Open Source Software Culture in the Undergraduate Computer Science Curriculum." *ACM SIGCSE Bulletin 39* (2): pp. 70–4. doi: 10.1145/1272848.1272888.

Dreisbach, J., T. Hogan, M. Stamford, and J. Greggo. 1998. "Focus Groups and Exit Interviews are Components of Chemistry Department Program Assessment." *Journal of Chemistry Education 75* (10): pp. 1330–2. doi: 10.1021/ed075p1330.

Drummond, T., and K. Owens. 2010. "Capturing Students' Learning." In *Engaging Student Voices in the Scholarship of Teaching and Learning*, edited by C. Werder and M. Otis, pp. 162–84. Sterling: Stylus.

Ericsson, K. A., and H. A. Simon. 1993. *Protocol Analysis: Verbal Reports as Data.* Cambridge: MIT Press.

Gibau, G. 2015. "Considering Student Voices: Examining the Experiences of Underrepresented Students in Intervention Programs." *CBE Life Sciences Education 14* (3), Article 28. doi: 10.1187/cbe.14-06-0103.

Herman, G., C. Zilles, and M. Loui. 2012. "Flip-Flops in Students' Conceptions of State." *IEEE Transactions on Education 55* (1): pp. 88–98.

Herrington, D., and P. Dabuenmire. 2014. "Using Interviews in CER Projects: Options, Considerations, and Limitations." In *Tools of Chemistry Education Research*, edited by D. Bunce and R. Cole, pp. 31–59. Washington: American Chemical Society.

Kearney, K., R. Damron, and S. Sohoni. 2015. "Observing Engineering Student Teams from the Organization Behavior Perspective Using Linguistic Analysis of Student Reflections and Focus Group Interviews." *Advances in Engineering Education 4* (3). Accessed August 30, 2017. http://advances.asee.org/?publication=observing-engineering-student-teams-from-the-organization-behavior-perspective-using-linguistic-analysis-of-student-reflections-and-focus-group-interviews.

Krueger, R., and M. Casey. 2014. *Focus Groups: A Practical Guide for Applied Research*, 5th ed. Thousand Oaks: Sage.

Litzinger, T., P. Van Meter, C. Firetto, L. Passmore, C. Masters, S. Turns,..., S. Zappe. 2010. "A Cognitive Study of Problem Solving in Statics." *Journal of Engineering Education 99* (4): pp. 337–53. doi: 10.1002/j.2168-9830.2010.tb01067.x.

McCracken, W., and W. Newstetter. 2001. "Text to Diagram to Symbol: Representational Transformation in Problem-Solving." *Proceedings of the 31st Annual ASEE/IEEE Frontiers in Education Conference*, Vol. 2: pp. F2G 13–17. Reno, NV: IEEE. doi: 10.1109/FIE.2001.963721.

Millis, B. 2004. "A Versatile Interactive Focus Group Protocol for Qualitative Assessment." In *To Improve the Academy*, Vol. 21, edited by C. M. Wehlburg and S. Chadwick-Blossey, pp. 125–41. San Francisco: Jossey Bass.

Mostyn, A., C. Jenkinson, D. McCormick, O. Meade, and J. Lymn. 2013. "An Exploration of Student Experiences of Using Biology Podcasts in Nursing Training." *BMC Medical Education 13*, Article 12. doi: 10.1186/1472-6920-13-12.

Partridge, L. and S. Sandover. 2010. "Beyond 'Listening' to the Student Voice: The Undergraduate Researcher's Contribution to the Enhancement of Teaching and Learning." *Journal of University Teaching & Learning Practice 7* (2), Article 4. Accessed August 30, 2017. http://ro.uow.edu.au/jutlp/vol7/iss2/4.

Patton, M. 2002. *Qualitative Research and Evaluation Methods*, 3rd ed. Thousand Oaks: Sage Publications, Inc.

Pressley, M. and P. Afflerbach. 1995. *Verbal Protocols of Reading: The Nature of Constructively Responsive Reading*. Hillsdale: Lawrence Erlbaum Associates, Publishers.

Prevost, L., and P. Lemons. 2016. "Step by Step: Biology Undergraduates' Problem-Solving Procedures During Multiple-Choice Assessment." *CBE Life Sciences Education 15* (4), Article 71. doi: 10.1187/cbe.15-12-0255.

Seidman, I. 2013. *Interviewing as Qualitative Research*, 4th ed. New York: Teachers College Press.

Shavelson, R., and L. Huang. 2003. "Responding Responsibly to the Frenzy to Assess Learning in Higher Education." *Change 35* (1): pp. 10–19.

Southard, K., T. Wince, S. Meddleton, and M. S. Bolger. 2016. "Features of Knowledge Building in Biology: Understanding Undergraduate Students' Ideas about Molecular Mechanisms." *CBE Life Sciences Education 15* (1), Article 7. doi: 10.1187/cbe.15-05-0114.

Trytten, D., A. Lowe, and S. Walden. 2012. "'Asians are Good at Math. What an Awful Stereotype': The Model Minority Stereotype's Impact on Asian American Engineering Students." *Journal of Engineering Education 101* (3): pp. 439–68.

Turner, D. W. 2010. "Qualitative Interview Design: A Practical Guide for Novice Investigators." *The Qualitative Report 15* (3): pp. 754–60. Accessed August 30, 2017. http://nsuworks.nova.edu/tqr/vol15/iss3/19.

Van Meter, P., M. Litzinger, M. Wright, and J. Kulikowich. 2006. "A Cognitive Study of Modeling During Problem Solving." Paper presented to the 2006 ASEE Annual Conference & Exposition, Chicago, Illinois, June 18–21. Accessed August 31, 2017. https://peer.asee.org/211.

van Someren, M., Y. Barnard, and J. Sandberg. 1994. *The Think Aloud Method: A Practical Guide to Modelling Cognitive Processes*. London: Academic Press. Accessed August 30, 2017. http://echo.iat.sfu.ca/library/vanSomeren_94_think_aloud_method.pdf.

Walker, D., M. Stremler, J. Johnston, D. Bruff, and S. Brophy. 2008. "Case Study on the Perception of Learning When Tablet PCs Are Used as a Presentation Medium in Engineering Classrooms." *International Journal of Engineering Education 24* (3): pp. 606–15.

Weinrich, M. L., and V. Talanquer. 2015. "Mapping Students' Conceptual Modes When Thinking About Chemical Reactions Used to Make a Desired Product." *Chemistry Education Research and Practice 16* (3): pp. 561–77. doi: 10.1039/C5RP00024F.

Weinrich, M. L., and V. Talanquer. 2016. "Mapping Students' Modes of Reasoning when Thinking about Chemical Reactions Used to Make a Desired Product."

Chemistry Education Research and Practice 17 (2): pp. 394–406. doi: 10.1039/C5RP00208G.

Werder, C., and M. Otis, editors. 2010. *Engaging Student Voices in the Scholarship of Teaching and Learning.* Sterling: Stylus.

Wright, L., J. Fisk, and D. Newman. 2014. "DNA → RNA: What Do Students Think the Arrow Means?" *CBE Life Sciences Education 13* (2): pp. 338–48. doi: 10.1187/cbe.CBE-13-09-0188.

CHAPTER 7

Analyzing Evidence

Introduction

Discussing specific statistical techniques for analyzing quantitative evidence is beyond the scope of this book. Investigators with sufficient background can consult texts or online resources such as Brase and Brase (2013), Gravetter and Wallnau (2011), Ramsey and Schafer (2013), or Trochim (2006). Most statistical tests can be performed using Excel, so specialized software such as SPSS, SAS, or R is not required. However, Tang and Ji (2014) have argued that there are reasons to learn and use R, a free and powerful open source statistical programming language. If researchers need to use more advanced statistical methods such as **factor analysis**, they may find it helpful to seek out a collaborator with statistical expertise, perhaps in another discipline.

Science, technology, engineering, and mathematics (STEM) instructors who want to embark on a scholarship of teaching and learning (SoTL) project but lack a background in statistics need not be too concerned. Many worthwhile projects, especially those belonging to the *What is?* category of the SoTL taxonomy, require no inferential statistics. Examples of such studies in engineering, computer science, and mathematics, respectively, can be found in Walker et al. (2008), Fulton and Schweitzer (2011), and Bennett and Dewar (2015).

In Chapter 4 (Quantitative Versus Qualitative Data) we noted that quantitative data may be a better fit for *What works?* questions and qualitative data for *What is?* questions, but it is possible for a *What works?* investigation to rely entirely on qualitative evidence (Gibau 2015). As Gibau observed, certain types of qualitative studies allow "for close examination of what types of interventions may or may not work at a given institution."

On the other hand, some quantitative studies not only seek to show an intervention is effective but also try to determine for whom it is most effective and what are the critical elements that make it effective. The latter are essentially *What is?* questions. For example, Eddy and Hogan (2014) used quantitative data to test the transferability of an increased course structure intervention (Freeman et al. 2011) across three contexts: (1) different instructors, (2) different student bodies (predominantly white and Asian vs. more diverse), and (3) different courses (majors vs. nonmajors). Statistical analyses were performed on total exam points earned and course and exam failure rates to determine whether an increased course structure intervention would transfer to a novel environment. Self-reported data on student race, ethnicity, national origin, and student generational status (first-generation vs. continuing generation college student) were collected using an in-class survey at the end of the term. Using Likert scale items, the survey also asked how much time students spent each week on various course activities, to what degree they perceived the class as a community, how frequently they attended, and how much they valued the skills learned. This survey data combined with the scores and failure rates enabled further statistical analyses to determine (1) if the effectiveness the intervention varied across different student populations, and (2) what factors might have influenced student achievement in the course with increased structure.

The remainder of this chapter focuses on methods for analyzing qualitative data followed by a comparison of the standards used to assess quantitative and qualitative research (Assessing the Quality of Content Analysis).

The analysis of qualitative data frequently involves **rubrics** or **coding**, neither of which may be familiar to STEM instructors. Rubrics are applied to student work such as exams, solutions, lab reports, papers, projects, or portfolios assigned in a course, work that is submitted for a grade. The rubric may be an integral part of the grading process or it can be applied separately to measure those aspects of learning that are the focus of the SoTL study. Coding is a method of analyzing textual data such as that gathered from interviews, focus groups, reflective writing assignments, or open-ended responses to surveys to determine common themes. Coding schemes can also be applied to written or recorded work to determine and categorize common mistakes, false starts, or misunderstandings. We examine rubrics and coding data in more detail.

Rubrics

A rubric is a guide for evaluating certain **dimensions** or characteristics of student work. In a SoTL study, these would be the skills or aspects of the learning under investigation. For each dimension, different **levels of performance** are defined, labeled, and described. A well-crafted rubric provides the criteria by which a task or piece of work will be judged and presents a rating scale with **descriptors** that distinguish among the ratings in the scale. The descriptors clarify the type of work that will be assigned to each level of achievement and help users apply the rubric consistently over time. We begin with some very basic examples of rubrics for assessing problem solving. Table 7.1 displays a simple rubric, derived from Charles et al. (1987, 30) for grading problem-solving tasks along three dimensions: Understanding the problem, Planning a solution, and Getting an answer.

Rubrics allow a SoTL researcher to reveal specific skills or aspects of learning that an intervention is having an effect on. For example, if a SoTL inquiry in mathematics wanted to know whether presenting heuristic strategies, such as "solve a simpler problem," "collect data to look for a pattern," and "work backwards" (Pólya 1957), improved students' problem-solving ability, the rubric in Table 7.1 would help to show the effect the instruction was having, in a way that pre- and post-scores on a problem-solving test would not. For an example of a rubric similar to the one shown in Table 7.1 applied to a student work sample, see Emenaker (1999, 118).

Table 7.1 *A simple rubric for a problem-solving task derived from Charles et al. (1987, 30)*

Dimension	Level of Performance Ratings		
	0	1	2
Understanding the problem	Complete misunderstanding of the problem	Part of the problem misunderstood or misinterpreted	Complete understanding of the problem
Planning a solution	No attempt, or totally inappropriate plan	Partially correct plan based on part of the problem being interpreted correctly	Plan could have led to a correct solution if implemented properly
Getting an answer	No answer, or wrong answer based on an inappropriate plan	Copying error; computational error; partial answer for a problem with multiple answers	Correct answer, correctly labeled

Some rubrics combine dimensions and ratings together into a single scale; these are called **holistic-scale rubrics**. Table 7.2 contains an example of a holistic-scale rubric for mathematical problem solving derived from the California Assessment Program (1989, 30). The descriptors given for six levels of problem-solving performance, ranging from "Unable to begin effectively" to "Exemplary response," would apply with minor modifications to many STEM fields. For a different adaptation of the same rubric (California Assessment Program 1989, 53) with five performance levels, see Emenaker (1999, 119). Sometimes the word "rubric" is applied more broadly, to include simple yes or no checklists (Crannell 1999).

Table 7.2 A holistic-scale rubric for mathematical problem solving derived from the California Assessment Program (1989, 53)

Score	Descriptor
6	*Exemplary response*: Gives a complete response with a clear, coherent, unambiguous and elegant explanation; includes a clear and simplified diagram; communicates effectively to the identified audience; shows understanding of the open-ended problem's mathematical ideas and processes; identifies all the important elements of the problem; may include examples and counterexamples; presents strong supporting arguments.
5	*Competent response*: Gives a fairly complete response with reasonably clear explanations; may include an appropriate diagram; communicates effectively to the identified audience; shows understanding of the problem's mathematical ideas and processes; identifies the most important elements of the problem; presents solid supporting arguments.
4	*Minor flaws but satisfactory*: Completes the problem satisfactorily, but the explanation may be muddled; argumentation may be incomplete; diagram may be inappropriate or unclear; understands the underlying mathematical ideas; uses mathematical ideas effectively.
3	*Serious flaws but nearly satisfactory*: Begins the problem appropriately but may fail to complete or may omit significant parts of the problem; may fail to show full understanding of mathematical ideas and processes; may make major computational errors; may misuse or fail to use mathematical terms; response may reflect an inappropriate strategy for solving the problem.
2	*Begins, but fails to complete problem*: Explanation is not understandable; diagram may be unclear; shows no understanding of the problem situation; may make major computational errors.
1	*Unable to begin effectively*: Words do not reflect the problem; drawings misrepresent the problem situation; copies parts of the problem but without attempting a solution; fails to indicate which information is appropriate to the problem.

An Internet search may turn up an existing rubric that can be used or modified. A project by the Association of American Colleges and Universities, *VALUE: Valid Assessment of Learning in Undergraduate Education*, developed 15 rubrics during an 18-month iterative process, including rubrics for written communication, oral communication, quantitative literacy, teamwork, and problem solving. It may be instructive to look at them (see Rhodes 2010). Another option, which we now examine, is to create one specifically for the investigation.

Basic Steps for Creating a Rubric

Rubrics can be developed for virtually any student work product, presentation, or behavior. As we have indicated, they make possible the assessment of separate aspects of a complex task. Because a rubric with dimensions and performance levels that align well with the research question can be a valuable tool in a SoTL study, we describe and then elaborate on the steps involved in creating one.

Step 1. Specify both the student learning goal that the rubric will be used to assess and the task or piece of work that it will judge.

Step 2. Identify the dimensions of the task or work that the rubric will measure and determine what each includes.

Step 3. For each dimension, decide on the number of performance levels and provide a description for each level. The number of performance levels can vary by dimension, and different weights can be given to the dimensions.

Step 4. Test the rubric and, if needed, revise or further develop it.

We now illustrate this process of developing a rubric for a fictitious SoTL study in a senior capstone STEM course in which students undertake research projects and give talks on their work. Suppose we plan to pair students as peer coaches for each other's end-of-semester presentation in an attempt to improve their oral presentations. The student learning outcome we wish to measure is how well students have developed the ability to give a STEM-discipline talk and we intend to apply a rubric to the end-of-semester presentation they make on their semester-long research project. That completes Step 1.

In Step 2 of creating the rubric, we identify four dimensions of the oral presentation that we want to measure, and we provide a detailed description (shown in parentheses) for each dimension:

- Preparation (speaker arrived on time with handouts or materials prepared, seemed familiar with the audio-visual projection system, presented a professional appearance, appeared to have practiced the talk, stayed within the time frame).

- Organization (talk had an introduction that captured the audience's attention and provided basic information with definitions or examples, as appropriate; a middle that presented data, summarized the method, argument, or proof, and gave the results obtained; and a clear end).
- Knowledge (speaker appeared to understand the underlying disciplinary content, was able to explain it clearly without constantly referring to notes, and could answer questions).
- Presentation Skills (presenter spoke clearly, and made eye contact).

For Step 3, we decide we will start with three performance levels. We can add more later, if finer distinctions are needed. There are many ways to label the levels, for instance, Poor-Acceptable-Good, Emerging-Developing-Advanced, or Limited-Acceptable-Proficient. Numerical values can be 1–2–3, 1–3–5, or something else. We can include sub-dimensions that are rated individually using the same or different scales. For example, in our Presentation Skills dimension, for the two sub-dimensions related to speaking clearly and maintaining eye contact we could use very simple descriptors: Never or rarely (with a value of 0 points), Some of the time (1 point), and Most or all of the time (2 points). If we choose the labels Needs Improvement, Adequate, and Very Good for the performance levels, we could represent this portion of the rubric by the chart shown in Table 7.3.

On the other hand, the more complex Preparation dimension needs more elaborate descriptors. Keeping the same labels—Needs Improvement, Adequate, and Very Good—we select 1–3–5 for numerical ratings. This will allow the possibility of assigning intermediate values 2 or 4 in performances that fall between the descriptors. The descriptors shown in Table 7.4 are based on our detailed description for the Preparation dimension. A zero (0) value could be assigned for a case where all five indicators of lack of preparation are present.

Table 7.3 *Rubric for the presentation skills dimension of giving a STEM-discipline talk*

Presentation Skills Sub-dimension	Level of Performance Ratings		
	Needs Improvement (0)	Adequate (1)	Very Good (2)
Eye contact is made	Never or rarely	Some of the time	Most or all of the time
Speaks clearly	Never or rarely	Some of the time	Most or all of the time

Table 7.4 *Rubric for the preparation dimension of giving a STEM-discipline talk*

Dimension	Level of Performance Ratings		
	Needs Improvement (1)	Adequate (3)	Very Good (5)
Preparation	Lack of preparation indicated by at least two of the following: confusion at the beginning over materials leads to delayed start; speaker struggles with projection system or presentation software due to lack of familiarity or incompatibility of equipment; length of talk departs from time allotted by a significant amount; speaker reads most of talk from notes; unprofessional appearance or demeanor detracts from speaker's credibility.	Talk starts on time; little, if any, problem with the projection system that could have been detected in advance; talk ends approximately on time; speaker is able to give talk with only occasional reference to notes; dress or demeanor does not detract from speaker's credibility.	Talk starts on time; no problem with the projection system that could have been detected in advance; talk is well-paced to end on time; speaker is able to give talk with little, if any, reference to notes; dress and demeanor enhance speaker's credibility.

To complete the rubric, we would need to construct similar tables for the other two dimensions: Organization and Knowledge. The Presentation Skills dimension has a total value of 4 points, while the Preparation dimension has a total of 5 points. If we feel some dimensions are more important than others, we could assign different weights to the dimensions.

For the final step, testing our rubric, we could try it on talks given in other courses, video recorded talks on the Internet, or on colloquium speakers who come to the department. We would ask students to comment on how clear and understandable the draft rubric was as a guide for their work. We would recruit colleagues to review or pilot test it and give feedback. Gathering input from other faculty members or students at earlier steps in the development of the rubric is also worth considering.

Applying a Rubric

If several people are going to use the same rubric in a given study, it is essential to practice applying it to the same presentation and discuss the results to arrive at a common understanding so that it can be employed consistently and fairly. This is called **norming a rubric**. Without this norming, raters may vary in their interpretations of the criteria as well as how rigorously to apply them. The process requires having student work samples that display different levels of performance on the various criteria (that is, the samples should not be all good nor all bad). The raters are given an orientation to the rubric that includes a discussion of student learning goal that the rubric is intended to assess, the dimensions that the rubric will measure, the performance levels, and the description for each level. All raters then concurrently and independently score the same samples, starting with the most straightforward one. Scores are compared, and consistencies and inconsistencies discussed, with the goal of articulating the reasons for the inconsistencies. The scoring criteria and descriptions are reviewed to find a way to revise them to produce agreement and develop a shared understanding of how to apply the rubric. This process is repeated with additional samples until the rubric is deemed sufficiently improved to minimize disagreement and rater bias.

Two simple measures of **inter-rater agreement** are the percentage of exact agreement and the percentage of exact and adjacent agreement (that is, the percentage of times ratings fall within one performance level of one another). The Standards for Educational and Psychological Testing (AERA et al. 1999) do not recommend a specific criterion for agreement, but they do state that an appropriate measurement should be calculated and reported. According to Graham et al. (2012), if there are five or fewer rating levels in the rubric, values from 75 percent to 90 percent demonstrate an acceptable level of agreement when using percentage of absolute agreement, but no ratings should be more than one level apart. Consult Graham et al. (2012) for additional details.

When a rubric is applied to student work, in this case to qualitative data in the form of oral presentation of a research project, it enables us to assess the performance along several dimensions, and produces a numerical rating for each dimension and for the presentation as a whole. Applying the rubric to presentations given in a senior capstone course, first in a semester when no peer coaching system was in place and later when there was one, would generate data on the efficacy of the peer coaching. We might find that some, but not all, aspects of the presentation were improved by peer coaching. Of course, evaluation by the rubric should not be the only data gathered for our hypothetical study. We might conduct a focus group or use interviews to obtain students' opinions of their peer coaching experience. Also, whether the students have access to the rubric prior to the presentation should be considered, because rubrics can have significant pedagogical value. They clarify the instructor's expectations for an assignment or project, allowing students to assess their own work. However, researchers should consider whether giving students advance access to rubrics that will be used to generate data for a SoTL study will skew the data.

For a more in-depth discussion of rubrics, consult *Introduction to Rubrics* (Stevens and Levi 2013).

Examples of STEM Education Studies that Used or Developed Rubrics

Addy and Stevenson (2014) used a rubric to assess the effects of critical thinking exercises that involved students reading and evaluating in writing biological claims made in the media or scientific publications. The rubric, developed by their institution Quinnipiac University, was based on the Association of American Colleges and Universities' *VALUE* rubric for Critical Thinking (Association of American Colleges and Universities 2009). It evaluates four dimensions of critical thinking (Ability to identify the problem, Assess the evidence provided, Take a position, and Form a conclusion) across four performance levels (Ineffective, Effective, Advanced, Outstanding). For example, the descriptor for the Advanced level of the Take-a-position dimension is "Specific position takes into account complexities of an issue. Other viewpoints are acknowledged; begins to synthesize other viewpoints with position" (Addy and Stevenson 2014, Appendix 4, 4).

A multidisciplinary group of faculty, including one biologist, examined how to improve students' critical reading skills for academic purposes and for social engagement (Manarin et al. 2015). Using elements from the Association for American Colleges and Universities' *VALUE* rubrics (Rhodes 2010), they developed and shared reading rubrics across their quite different disciplines (English, political science, biology, and history).

A group consisting of seven faculty members from engineering (electrical, chemical, and mechanical), engineering education, and management, and two consultants (with backgrounds in mathematics and structural engineering) collaborated to develop and validate a web-based instrument for self- and peer-evaluation of teamwork (Ohland et al. 2012). The instrument, a rubric, measures team-member contributions in five dimensions that were chosen based on the team effectiveness literature (Contributing to the work, Interacting with team members, Keeping the team on track, Expecting quality, Having relevant knowledge, skills, and abilities). Each of five performance levels is associated with certain behaviors. For example, the mid-level performance on interacting with team members is described as follows:

- Listens to teammates and respects their contributions.
- Communicates clearly. Shares information with teammates. Participates fully in team activities.
- Respects and responds to feedback from teammates (Ohland et al. 2012, 626).

Our final example of a rubric also stems from a think-aloud. To assess the performance of students solving problems in statics during a think-aloud, Litzinger and his colleagues (2010) first consulted the literature to develop their Integrated Problem-Solving Model. Then the model provided the basis for the rubric they used to evaluate and code participants' responses. In what we consider testimony to the value of rubrics for SoTL studies, they commented, "the findings of this study not only advance our understanding of the sources that underlie the problem-solving struggles of statics students, but also suggest a means for intervening to address these struggles" (349).

> **Pause: Analyzing Evidence with Rubrics—Questions to Consider**
>
> - Will a rubric facilitate the analysis of the evidence I collect?
> - Is there an appropriate rubric available, one that is aligned with some aspect of my question?
> - If I am going to create a rubric, what literature might I consult?
> - In creating my rubric, have I determined the dimensions to be evaluated and levels of performance and written a clear description for each pair of dimensions and performance levels?
> - Have I tested the rubric either through a pilot study or small sample?
> - If more than one person will be applying the rubric, what steps will be taken to standardize (norm) the use of it?
> - Will students be given access to the rubric before they complete the assignment or work that it will be applied to? If so, have I considered how that might affect the research results?

Coding Data (Content Analysis)

We now turn our attention to a method to distill and categorize information from verbal data such as that found in transcripts, reflective writing assignments, or open-ended responses to survey questions. Data collection methods such as think-alouds, focus groups, and interviews also produce verbal data, typically a lot of it. Meaning is extracted from this data by using labels (codes) to identify **common themes** that reoccur throughout a data set. The codes make it possible to categorize and organize information as it relates to the research question of the SoTL study. This **content analysis** allows the researcher to discover patterns that would be difficult to detect by reading alone. The resulting codes help the SoTL scholars to communicate findings to others. The biggest concern when a researcher is new to **coding data** is how to find the right interpretation. But, in practice, there can be many right interpretations. What is essential is to describe the coding process that was used and then explain whatever resulting interpretation is put forth, typically by providing textual samples, so that others can understand the results and how they were obtained (Auerbach and Silverstein 2003, 32).

Reading the raw text, let alone analyzing it and drawing conclusions about patterns, can seem an overwhelming task. We begin with an overview and several examples before we outline the steps in the coding process.

Predetermined Categories

The categories or codes that we use can be "predetermined" or allowed to "emerge" from the data. **Predetermined categories**, sometimes called **a priori categories**, are those suggested by the research question or a published theory. For example, in mathematics, a research question might be: "Does using graphing calculators cause students to think more geometrically?" Students could be given similar mathematical tasks to perform first without and then with graphing calculators, or before and after using graphing calculators as a tool. The investigation might use a think-aloud methodology. The data could be analyzed and categorized according to whether it contains markers of *geometric* or *algebraic* thinking.

Derek Bruff (2015) describes two coding schemes that he developed, both based on the revised Bloom's taxonomy (Anderson and Krathwohl 2001). He used the six categories—Remember, Understand, Apply, Analyze, Evaluate, and Create—from the taxonomy's cognitive process dimension to code the questions he had asked students on pre-reading quizzes. To analyze and describe the difficulties his students reported having, he coded their responses to this "muddiest-point" question: "Please write about the part of the reading that you found most difficult. Try to be as specific as possible. If you did not find the reading difficult, write about the part you found most interesting" (130). This time his codes—Factual, Conceptual, Procedural, and Metacognitive—were derived from the knowledge dimension of Anderson and Krathwohl's (2001) revised Bloom's taxonomy.

In both examples, the researchers began coding knowing, or at least thinking they knew, exactly what to look for. In our first example, it was markers of geometric thinking versus indications of algebraic approaches. In the second, Bruff relied on categories drawn from a well-known taxonomy.

Predetermined categories can be as simple as Positive, Negative, and Neutral. This coding scheme sufficed to ascertain students' perceptions (obtained from their responses to an open-ended survey question) of the effectiveness of working in groups on realistic projects to learn statistics or to develop communication skills in Holcomb (2015).

Emergent Categories

Sometimes the researcher has no idea what categories or themes to look for and has to let the data suggest the themes. This is called using **emergent categories** or **inductive coding**. A study of students' views of mathematics provides an example of this type of coding (Dewar 2008). The study investigated how undergraduate STEM students' understandings of mathematics compared to an expert view of mathematics, and whether a single course could enhance future teachers' views of mathematics. Written responses to the question "What is mathematics?" were gathered first from 55 mathematics and computer science students and 16 mathematics faculty members, and a few years later from seven future teachers at the beginning and end of a particular course. Dewar and a colleague (Bennett) independently read the original set of responses several times and were able to identify many obvious and repeated themes, such as "mathematics was about numbers" or it "involved making logical arguments." After comparing notes, they settled upon six emergent categories and provided an example for each category, as shown in Table 7.5. When yet another colleague was able to use the information in Table 7.5 to code the original data set with similar results, Dewar was confident that others could understand the coding scheme. These emergent categories then became predetermined categories for coding the responses of the seven future teachers.

After completing the coding, Dewar found the work of Schwab (1964) on classifying the disciplines. She noticed that the six emergent categories each aligned with one of the four bases used by Schwab to classify disciplines—Content boundaries, Skills and habits employed by practitioners, Modes of inquiry, and Purposes or outcomes for the disciplinary work. The alignment of the categories with Schwab's theoretical framework provided theoretical support for the appropriateness of the emergent categories. This example shows how a return to the literature can benefit a study.

Table 7.5 *Emergent categories and example responses to* **What is mathematics?**

Emergent Category	Example Response
Numbers (including computation)	the study of numbers
Listing of Topics (list could include numbers)	algebra, pictures, numbers, everything encompasses math
Applications	a "mental gymnastic" that helps you solve real-world and theoretical problems
Pattern, Proof, or Logic	the search for and the study of patterns
Structure, Abstraction, or Generalization	the analysis of abstract systems
Other	a language

Process for Coding Data

We now present a step-by-step process for coding, derived from Auerbach and Silverstein (2003), Taylor-Powell and Renner (2003), and the Visible Knowledge Project (2009). The process is reasonable for coding ten to 50 pages of text manually. For larger data sets, consider using specialized software that is available for the content analysis of qualitative data.

1. Put the data into a single (Word or Excel) file. Assign an identifying name or number to each respondent's data.
2. Read and reread the file to become familiar with the data.
3. Keeping the research question in mind, select segments of text that address it.
4. Label in the margin why the segment of text was selected either by hand or by using Word's comment feature. This is called **memo-ing**. The codes or categories will begin to emerge. For a simple example, see *Analyzing Qualitative Data* (Taylor-Powell and Renner 2003, 4).
5. Keep a master list of labels (codes) generated in Step 4.

That completes the **initial coding**. The next phase, **focused coding**, includes the steps:

6. Review, revise, and reorganize the codes. Eliminate less useful codes, combine smaller categories into larger ones, and subdivide categories with a large number of segments.
7. Examine the codes for related ideas to identify larger themes or to develop theory.
8. For advanced work, the final step would be to build a visual representation of the data via a hierarchical diagram, a relations chart, or a network diagram.

In our discussion (see Coding Data, Emergent Categories) of the study of students' views of mathematics (Dewar 2008), we noted that she worked collaboratively with colleagues first to develop and later to check the coding system. Although not listed explicitly as one of the steps, an important part of the coding process is to provide an "external check on the highly interpretive coding process" (Creswell 2013, 253) by demonstrating that multiple coders can obtain the same results. This is known as **inter-coder reliability.** There are many ways of going about this process, and it may involve seeking agreement on what code names are used, which passages are coded, and whether the same passages are coded in the same way. Often the process begins once the researcher has settled upon a set of code names, short descriptors for each code name, and an example phrase or response for each of the coding categories. Then one or more colleagues are selected to code some or all of the data based on the code names, descriptors, and exemplars. Miles and Huberman (1994, 64) recommend pre-testing and revising coding schemes, because initial coding instructions often yield poor agreement. Eighty percent agreement is a frequently mentioned goal (Creswell 2013). The agreement level can be calculated by percent of agreement between two coders, by averaging the percent of pairwise agreement between multiple coders, or by more complicated statistics such as Krippendorff's alpha (Krippendorff 2004). Depending on the statistic used, the acceptable level of agreement may vary, but the value is usually between 70 percent and 90 percent (Guest and MacQueen 2008, 251). Often conversations among coders aimed at achieving consensus about the codes that differ are required to arrive at a sufficient level of agreement. If the data set is large, a sufficient level of consensus may be demonstrated on just a portion of the data (see, for example, the study by Weaver et al. (2009) described in the following section (Examples of STEM Education Studies that Used Coding)). Whatever approach is used should be reported when communicating the results.

Examples of STEM Education Studies that Used Coding

Coding using both emergent and predetermined categories took place in an exploratory study of how scientific discourse in an online environment would differ from that occurring in a face-to-face (F2F) situation (Weaver et al. 2009). Student-instructor interactions in both the online and the F2F environments took the form of questions and responses. The questions were coded for question type (Clarification, Class content, Prompting, Leading, Technology, Class administration) and level (Input, Processing, and Output). The question type indicates its intent or purpose, which the researchers determined by reading the question in context (hence these are emergent categories). The question level is associated with the degree of mental work required to answer the question. These categories were based on a taxonomy developed by Costa (1985). Additional predetermined codes (Student responses, Student questions, Instructor responses, and Instructor questions) represented the participant who posed the statement or question. Each of these was further divided based on content and gender.

Weaver et al. (2009) reported their efforts to achieve an acceptable level of inter-coder reliability as follows:

> Two researchers independently coded a set of five transcripts, and discrepancies between researchers were counted in order to calculate inter-rater reliability using percent agreement. Discrepancies were discussed for the sake of improving agreement between researchers, and this cycle was repeated for subsequent sets of five transcripts until inter-rater reliability was found to remain above 90% for two successive sets of transcripts. After this point, the remaining transcripts were coded by one researcher only (7).

In their study, Irving and Sayre (2013) sought descriptions of categories of influence based on the students' own experiences. Therefore, they rejected the use of previously identified elements of physics students' identity development and allowed the categories to emerge from the data. Three stages of identity development resulted from the semi-structured interviews they conducted with 30 upper-level physics students: Student, Aspiring physicist, and Physicist. They found that variation in three factors—students' career definitiveness, their metacognitive level, and their description of when one has become a physicist—would distinguish among the three stages of identity development. Their content analysis process is discussed in detail in their article (Irving and Sayre 2013, 72). What follows summarizes just a few of the many steps they describe taking. Each transcript was read repeatedly, often in one sitting, in order to become acquainted with the transcript set as a whole. For each reading, the focus was on one particular aspect. For example, on one occasion, the focus may have been on how the students described their first experiences with physics, on another occasion, attention would be paid to those aspects of physics that the students described as liking, and on yet another, the focus would be on students' conceptions of understanding. Preliminary categories were determined and then reexamined to ensure that the categories faithfully and empirically described the variations in the stages of identity development of this particular group of students. The last step was to place each individual student into one of the three categories. In one case of disagreement, the researchers negotiated to determine the appropriate placement of a student.

Coding Non-verbal Data

Non-numerical, non-verbal data such as graphs, diagrams, or concept maps can also be coded. To do this, the researcher would identify characteristics of interest and describe them in words. Williams (1998) and Kinchin et al. (2000) are two examples of studies that involved coding concept maps. The first study employed emergent coding on concept maps to examine differences between students in reform and traditional calculus courses in their conceptual understanding of functions. The second study described a coding scheme based on the structure of the maps generated by Year 8 students while reviewing the reproduction of flowering plants. The researchers observed that the map structures that had emerged in this setting were seen in subsequent studies conducted with students at other levels (including postgraduate students) involving different topics in the biological sciences. This is an example of transferability of a coding scheme to other populations and situations. For a further discussion of transferability, see the section Assessing the Quality of Content Analysis in this chapter.

> **Pause: Analyzing Evidence by Coding—Questions to Consider**
> - Will the qualitative evidence I collect be analyzed by coding?
> - Will the data I collect need to be transcribed?
> - Will I code by hand or use software?
> - Have I identified a set of predetermined categories that I could use in my coding? Is it well-aligned to some aspect of my question?
> - Will I use a coding scheme that emerges from the data?
> - What measure and level of inter-coder reliability will I use?
> - How will I ensure there is inter-coder reliability?

Assessing the Quality of Content Analysis

We begin with a summary of how one measures the quality of quantitative research. This will serve as a point of comparison to help us explicate standards for qualitative research. By quantitative research we mean the use of experimental methods and quantitative measures to test hypothetical generalizations. The phenomenon under study is regarded as consisting of observable facts that can be measured by some instrument. The excellence of quantitative research is judged by the **reliability** and **validity** of the instruments and the **generalizability** of the results. If a finding from a study conducted on a sample population can be applied to the population at large, the result is said to be generalizable. A well-designed, well-executed study that uses the gold standard of random assignment of subjects to experimental and control groups is considered to yield generalizable results. In Chapter 3 (Challenges of Education Research Design for SoTL) we discussed why this may be an unattainable goal for education research.

In Chapter 5 (Surveys, Reliability and Validity) we discussed reliability and validity in the context of survey instruments. In general, a reliable instrument will yield consistent results, while validity is the extent to which an instrument measures what it claims to measure (Reynolds et al. 2006). It is possible for an assessment tool to be consistent but lack validity.

What are the corresponding standards for qualitative research? According to Auerbach and Silverstein (2003, 77–87), qualitative data analysis must be **justifiable**. This means it has the following three characteristics:

1 The steps used to arrive at the interpretation are **transparent**. This does not mean others would end up with the same interpretation, only that they understand how the researcher obtained the interpretation.
2 The themes that were developed are **communicable**. This means that others can understand the themes even though they may not agree with them.
3 The themes (and constructed theories, if any) tell a **coherent** story.

Being justifiable is the qualitative researcher's counterpart to reliability and validity in quantitative analysis.

Another measure of quality is whether larger, more abstract themes (that is, constructed theories) are **transferable** to other populations or situations. Small SoTL studies usually don't involve developing larger constructs, so transferability may not come into play. Being transferable is the qualitative researcher's counterpart to generalizability in quantitative analysis. One additional measure of the quality of a content analysis of qualitative data is **inter-coder reliability**. As previously discussed (Coding Data, Process for Coding Data), this refers to others being able to code with similar results given categories, descriptors, and example segments of text. This stability of responses by multiple coders shows the reliability of the coding scheme.

The content analysis employed in the study of students' views of mathematics (Dewar 2008) discussed earlier in this chapter (see Coding Data, Emergent Categories) illustrated several of these concepts. Because the codes developed from STEM majors and mathematics faculty members could be applied to a group of future teachers, most of whom were preparing to teach elementary school (hence non-STEM majors), this study provided an example of transferability of the codes (but not a larger theory) to a different population. Being able to correlate the categories with the work of Schwab (1964) showed that they merged with a larger theory. That a colleague was able to code the original data set successfully based on the categories and examples in Table 7.5 provided evidence that the themes were communicable. A different SoTL study (Szydlik 2015) also analyzed student responses to the question "What is mathematics?" with similar, but not identical, categories emerging, thus demonstrating the possibility of more than one "right" interpretation.

> **Pause: Analyzing Evidence—Questions to Consider**
>
> - How will I analyze the quantitative evidence I gather?
> - How will I analyze the qualitative evidence I gather?
> - How will my quantitative analysis and results measure up against the standards of reliability, validity, and generalizability?
> - How will my qualitative analysis and results measure up against the standards of justifiability (which includes transparency, communicability, and coherence), inter-coder reliability, and transferability?

REFERENCES

Addy, T., and M. Stevenson. 2014. "Evaluating Biological Claims to Enhance Critical Thinking through Position Statements." *Journal of Microbiology and Biology Education* 15 (1): pp. 49–50. doi: 10.1128/jmbe.v15i1.663.

AERA, APA, and NCME (American Educational Research Association, American Psychological Association, and National Council on Measurement in Education). 1999. *Standards for Educational and Psychological Testing*. Washington: AERA.

Anderson, L., and D. Krathwohl. 2001. *Taxonomy for Learning, Teaching and Assessing: A Revision of Bloom's Taxonomy of Educational Objectives*. New York: Longman.

Association of American Colleges and Universities. 2009. *Critical thinking VALUE rubric*. Accessed August 31, 2017. http://www.aacu.org/value/rubrics/critical-thinking.

Auerbach, C., and L. Silverstein. 2003. *Qualitative Data: An Introduction to Coding and Analysis*. New York: New York University Press.

Bennett, C., and J. Dewar. 2015. "The Question of Transfer: Investigating How Mathematics Contributes to a Liberal Education." In *Doing the Scholarship of Teaching and Learning in Mathematics*, edited by J. Dewar and C. Bennett, pp. 183–90. Washington: Mathematical Association of America.

Brase, C. H., and C. P. Brase. 2013. *Understanding Basic Statistics*, 6th ed. Boston: Brooks/Cole.

Bruff, D. 2015. "Conceptual or Computational? Making Sense of Reading Questions in an Inverted Statistics Course." In *Doing the Scholarship of Teaching and Learning in Mathematics*, edited by J. Dewar and C. Bennett, pp. 127–36. Washington: Mathematical Association of America.

California Assessment Program. 1989. *A Question of Thinking: A First Look at Students' Performance on Open-Ended Questions in Mathematics*. Sacramento: California State Department of Education.

Charles, R., F. Lester, and P. O'Daffer. 1987. *How to Evaluate Progress in Problem Solving*. Reston: National Council of Teachers of Mathematics.

Costa, A. L. 1985. "Toward a Model of Human Intellectual Functioning." In *Developing Minds: A Resource Book for Teaching Thinking*, edited by A. L. Costa, pp. 62–5: Alexandria: Association for Supervision of Curriculum Development.

Crannell, A. 1999. "Assessing Expository Mathematics: Grading Journals, Essays, and Other Vagaries." In *Assessment Practices in Undergraduate Mathematics*, edited by B. Gold, S. Keith, and W. Marion, pp. 113–15, MAA Notes Series #49. Washington: Mathematical Association of America. Accessed August 30, 2017. http://www.maa.org/sites/default/files/pdf/ebooks/pdf/NTE49.pdf.

Creswell, J. 2013. *Qualitative Inquiry & Research Design: Choosing among Five Approaches*, 3rd ed. Thousand Oaks: Sage.

Dewar, J. 2008. "What Is Mathematics: Student and Faculty Views." *Electronic Proceedings for the Eleventh Special Interest Group of the Mathematical Association of America on Research in Undergraduate Mathematics Education Conference on Research in Undergraduate Mathematics Education*, San Diego, CA. Accessed August 30, 2017. http://sigmaa.maa.org/rume/crume2008/Proceedings/dewar_SHORT.pdf.

Eddy, S., and K. Hogan. 2014. "Getting under the Hood: How and for Whom Does Increasing Course Structure Work?" *CBE Life Sciences Education* 13 (3): pp. 453–68. doi: 10.1187/cbe.14-03-0050.

Emenaker, C. 1999. "Assessing Modeling Projects in Calculus and Precalculus: Two Approaches." In *Assessment Practices in Undergraduate Mathematics*, edited by B. Gold, S. Keith, and W. Marion, pp. 116–19. MAA Notes Series #49. Washington: Mathematical Association of America. Accessed August 30, 2017. http://www.maa.org/sites/default/files/pdf/ebooks/pdf/NTE49.pdf.

Freeman, S., D. Haak, and M. Wenderoth. 2011. "Increased Course Structure Improves Performance in Introductory Biology." *CBE Life Sciences Education* 10 (2): pp. 175–86. doi: 10.1187/cbe.10-08-0105.

Fulton, S., and D. Schweitzer. 2011. "Impact of Giving Students a Choice of Homework Assignments in an Introductory Computer Science Class." *International Journal for the Scholarship of Teaching and Learning* 5 (1). Accessed August 30, 2017. http://digitalcommons.georgiasouthern.edu/cgi/viewcontent.cgi?article=1270&context=ij-sotl.

Gibau, G. 2015. "Considering Student Voices: Examining the Experiences of Underrepresented Students in Intervention Programs." *CBE Life Sciences Education* 14 (3), Article 28. doi: 10.1187/cbe.14-06-0103.

Graham, M., A. Milanowski, and J. Miller. 2012. *Measuring and Promoting Inter-rater Agreement of Teacher and Principal Performance Ratings*. Center for Educator Compensation Reform. Accessed August 31, 2017. http://files.eric.ed.gov/fulltext/ED532068.pdf.

Gravetter, F., and L. Wallnau. 2011. *Essentials of Statistics for the Behavioral Sciences*, 7th ed. Belmont: Wadsworth/Cengage Learning.

Guest, G., and K., MacQueen, editors. 2008. *Handbook for Team-Based Qualitative Research*. Lanahm: AltaMira Press.

Holcomb, J. 2015. "Presenting Evidence for the Field that Invented the Randomized Clinical Trial." In *Doing the Scholarship of Teaching and Learning in Mathematics*, edited by J. Dewar, and C. Bennett, pp. 117–26. Washington: Mathematical Association of America.

Irving, P., and E. Sayre. 2013. "Physics Identity Development: A Snapshot of the Stages of Development of Upper-Level Physics Students." *Journal of the Scholarship of Teaching and Learning 13* (4): pp. 68–84. Accessed August 30, 2017. http://josotl.indiana.edu/article/view/3521.

Kinchin, I., D. Hay, and A. Adam. 2000. "How a Qualitative Approach to Concept Map Analysis Can Be Used to Aid Learning by Illustrating Patterns of Conceptual Development." *Educational Researcher* 42 (1): pp. 43–57. doi: 10.1080/001318800363908.

Krippendorff, K. 2004. *Content Analysis: An Introduction to Its Methodology*. Thousand Oaks: Sage.

Litzinger, T., P. Van Meter, C. Firetto, L. Passmore, C. Masters, S. Turns,...., Zappe, S. 2010. "A Cognitive Study of Problem Solving in Statics." *Journal of Engineering Education*, 99 (4): pp. 337–53. doi: 10.1002/j.2168-9830.2010.tb01067.x.

Manarin, K., M. Carey, M. Rathburn, and G. Ryland. 2015. *Critical Reading in Higher Education: Academic Goals and Social Engagement*. Bloomington: Indiana University Press.

Miles, M., and A. Huberman. 1994. *Qualitative Data Analysis: An Expanded Sourcebook*, 2nd ed. Thousand Oaks: Sage.

Ohland, M., M. Loughry, D. Woehr, L. Bullard, R. Felder, C. Finelli, ..., D. Schmucker. 2012. "The Comprehensive Assessment of Team Member Development of a Behaviorally Anchored Rating Scale for Self- and Peer Evaluation." *Academy of Management Learning & Education* 11 (4): pp. 609–30. doi: 10.S465/amle.2010.0177.

Pólya, G. 1957. *How to Solve It*, 2nd ed. Garden City: Doubleday.

Ramsey, F., and D. Schafer. 2013. *The Statistical Sleuth: A Course in Methods of Data Analysis*, 3rd ed. Boston: Brooks/Cole.

Reynolds, C., R. Livingston, and V. Wilson. 2006. *Measurement and Assessment in Education*. Boston: Pearson Education Inc.

Rhodes, T., editor. 2010. *Assessing Outcomes and Improving Achievement: Tips and Tools for Using Rubrics*. Washington: American Association of Colleges and Universities.

Schwab, J. 1964. "Structure of the Disciplines." In *The Structure of Knowledge and the Curriculum*, edited by G. W. Ford and L. Pugno, pp. 6–30. Skokie: Rand McNally.

Stevens, D., and A. Levi. 2013. *Introduction to Rubrics: An Assessment Tool to Save Grading Time, Convey Effective Feedback, and Promote Student Learning*, 2nd ed. Sterling: Stylus.

Szydlik, S. 2015. "Liberal Arts Mathematics Students' Beliefs about the Nature of Mathematics: A Case Study in Survey Research." In *Doing the Scholarship of Teaching and Learning in Mathematics*, edited by J. Dewar and C. Bennett, pp. 145–65. Washington: Mathematical Association of America.

Tang, H., and P. Ji. 2014. "Using the Statistical Program R instead of SPSS to Analyze Data." In *Tools of Chemistry Education Research*, edited by D. Bunce and R. Cole, pp. 135–51. Washington: American Chemical Society.

Taylor-Powell, E., and M. Renner. 2003. *Analyzing Qualitative Data*. Madison: University of Wisconsin Extension. Accessed August 31, 2017. https://learningstore.uwex.edu/assets/pdfs/g3658-12.pdf.

Trochim, W. 2006. "Inferential Statistics." *Research Methods Knowledge Base*. Updated October 20. http://www.socialresearchmethods.net/kb/statinf.php.

Visible Knowledge Project. 2009. "Coding Data: Methods & Collaboration." Blog. January 19. Accessed August 31, 2017. https://blogs.commons.georgetown.edu/vkp/2009/01/16/coding-data1.

Walker, D., M. Stremler, J. Johnston, D. Bruff, and S. Brophy. 2008. "Case Study on the Perception of Learning When Tablet PCs Are Used as a Presentation Medium in Engineering Classrooms." *International Journal of Engineering Education* 24 (3): pp. 606–15.

Weaver, G., K. Green, A. Rahman, and E. Epp. 2009. "An Investigation of Online and Face-to-Face Communication in General Chemistry." *International Journal for the Scholarship of Teaching and Learning* 3 (1), Article 18. doi: 10.20429/ijsotl.2009.030118.

Williams, C. 1998. "Using Concept Maps to Assess Conceptual Knowledge of Functions." *Journal for Research in Mathematics Education* 29 (4): pp. 414–21.

CHAPTER 8

The Final Step for Doing SoTL
Going Public

Introduction

The view of scholarship of teaching and learning (SoTL) that we present in this book calls for SoTL researchers to submit their findings to peer review and make them public for others in the academy to build upon. In higher education going public typically involves making conference presentations, possibly with publication in the conference proceedings, or publishing articles in scholarly journals, chapters in edited volumes, or entire books. This chapter gives advice and resources for completing a SoTL project and disseminating the results. These include why it is a good idea to find collaborators for doing SoTL and where to find them, sources of support, possible venues for dissemination, and advice for completing a manuscript for publication. In many instances, our suggestions for successfully completing SoTL projects parallel common practices in disseminating disciplinary research and may already be quite familiar to faculty with significant publishing experience.

Finding Collaborators and Support

SoTL lends itself to collaboration both formal and informal. During our Carnegie scholar years, we were placed into small interdisciplinary teams of six

The Scholarship of Teaching and Learning, Jacqueline M. Dewar, Curtis D. Bennett, and Matthew A. Fisher.
© Oxford University Press, 2018. Published 2018 by Oxford University Press

to eight scholars to be mentored by a Carnegie staff member or a scholar from a prior cohort. Within each team, each scholar was paired with another team member to act as **critical friends**, that is, peer mentors, to one another. Through this experience, we learned first-hand the value of having a colleague, even (or perhaps, especially) outside one's own discipline, to act as a sounding board and give feedback. Higgs (2009) described how the critical friends model of faculty development benefitted the Irish Integrative Learning Project by "[helping] participants to examine their work from another perspective, and to receive a critique from a colleague who is an advocate for the success of their work."

Tsang (2010, 4) cautioned against treating SoTL as a "solitary endeavor" and we agree. Collaborators provide new insights or access to expertise or strengths beyond that of a single scholar (Bennett and Dewar 2004), just as they do in traditional disciplinary research. Sawrey (2008) made similar observations for chemistry education research: "Collaboration can provide intellectual and monetary support for projects, and help broaden the audience for outcomes... [and] can lead to fruitful synergies, even without funding" (212).

Much of what we have to say about collaborators will parallel experiences in the academic disciplines. SoTL collaborations occur both within and across disciplines. For example, of the 15 SoTL projects in mathematics described in Dewar and Bennett (2015), six involved collaborative work and three of those included involved collaboration with non-mathematicians, including faculty from psychology, biology, and education, and staff from academic support services. Teaching or faculty development centers, institutional assessment units, and instructional technology departments are other places to find potential collaborators.

Potential off-campus collaborators may be found at conference sessions or workshops focused on science, technology, engineering, or mathematics (STEM) teaching, STEM education, SoTL, liberal education, civic engagement, assessment, technology use, or teacher preparation. SoTL researchers can look for presenters who are doing related work, using methods that might translate to their situation, or cite references that seem to apply to their study. The typical networking advice—to ask questions before or after potential collaborator's talks, get contact information, and follow up later by email—applies.

Professional organizations that promote and support SoTL include the International Society for the Exploration of Teaching and Learning (ISETL), the International Society for the Scholarship of Teaching and Learning (ISSOTL), and the Society for Teaching and Learning in Higher Education (STLHE). Each holds an annual conference where colleagues from many disciplines gather to discuss and present scholarly work related to teaching and learning. These multi-disciplinary gatherings afford good opportunities for meeting potential collaborators. The Internet addresses for several sites that maintain comprehensive lists of SoTL conferences are given in the Selecting a Conference section in this chapter.

Faculty members interested in SoTL work often ask about funding sources to support it. Many campuses provide small amounts of funding for course or curricular development, experimenting with innovative teaching methods, or incorporating technology into courses. These may not be labeled as SoTL grants, but including SoTL as part of the assessment plan should make an application more competitive.

In the United States (US), external funding for larger projects involving STEM teaching and learning may be available through the American Educational Research Association, the Association for Institutional Research, the Department of Education, the Environmental Education grants from the Environmental Protection Agency, Learn and Serve America, a program of the Corporation for National and Community Service, the National Science Foundation, and various funding opportunities administered by professional societies in STEM fields. Private foundations that offer funding to improve teaching and learning in STEM fields include the Ford Foundation, the Henry Luce Foundation, and the Spencer Foundation. Many campuses have an office to assist faculty members in seeking external funding that can help identify additional funding sources.

Other countries also have potential sources of funding for SoTL projects. In Australia, the Office for Learning and Teaching is the most significant source of funding for SoTL work. There are also a number of foundations in that country that fund education research work, either exclusively or among other priorities. Education funding in Canada is largely handled at the provincial level, so funding models and sources will depend on where a faculty member is located.

Presenting a Paper

Just as in disciplinary research, a conference paper or poster presentation provides an opportunity to interact with and obtain advice from others who are interested in the topic. Poster presentations allow for more informal one-on-one discussions. It varies by discipline and conference whether a full paper or only an abstract must be submitted for consideration prior to the conference.

Prior to submitting a journal article, presenting a preliminary report at a conference can be a good way to get feedback from others.

Selecting a Conference

Many of the larger professional societies in the STEM disciplines that hold annual conferences to disseminate research in the disciplines and to address professional concerns include strands or sessions for presenting educational research and SoTL. Some examples are the annual meetings of the American Chemical Society and the American Physical Society, and the annual joint meeting of the American Mathematical Society and the Mathematical Association of America. There are similar conference options provided by professional societies for the biological and other sciences and for engineering as well as meetings sponsored by professional societies outside the US.

Usually, these same STEM disciplines have other groups sponsoring conferences that are entirely focused on research and scholarship related to teaching and learning. In chemistry, the Division on Chemical Education of the American Chemical Society sponsors the Biennial Conference on Chemical Education. In mathematics, the Special Interest Group of the Mathematical Association of America on Research in Undergraduate Mathematics Education organizes an annual conference. The American Association of Physics Teachers organizes both summer and winter meetings each year. In biology, annual conferences are sponsored by the Society for the Advancement of Biology Education Research and the American Society for Microbiology. In engineering, the American Society for Engineering Education hosts an annual conference and other workshops and institutes for particular audiences (e.g., deans) or with specific themes (e.g., for industry and education collaboration). An Internet search specifying a particular STEM discipline will generate additional examples.

In addition to professional societies, other higher education organizations organize conferences with a focus on STEM education. For example, the Association of American Colleges and Universities sponsors an annual Transforming STEM Higher Education conference within its Network for Academic Renewal. Several conferences appropriate for presenting SoTL work are held in Australia. The conference of the Higher Education Research and Development Society of Australasia is probably the best known, but there is also the Australasian Science Education Research Association conference.

There are also conferences that focus on scholarship of teaching and learning across all disciplines. Among these are Lilly Conferences on College and

University Teaching and the Teaching Professor Conference in the US, the Symposium on Scholarship of Teaching and Learning sponsored by Mount Royal University in Canada, EuroSoTL in the European Union, and the annual ISSOTL conference, which rotates locations through Australia, North America, and Europe. The ISSOTL website (http://www.issotl.com) includes links to SoTL conferences held in different regions of the world.

The Center for Excellence in Teaching and Learning (CETL) at Kennesaw State University maintains one of the most comprehensive directories of teaching conferences at https://cetl.kennesaw.edu/teaching-conferences-directory. The list is searchable by a number of characteristics such as geographical location, discipline, and topic (including scholarship of teaching and learning) and includes links to the specific conference websites. The University of Washington Center for Teaching and Learning and the Office of the Cross Chair of Scholarship of Teaching and Learning at Illinois State University also maintain lists of upcoming SoTL conferences at http://www.washington.edu/teaching/sotl-annual-conferences and http://sotl.illinoisstate.edu/conferences, respectively.

Additional factors to take into account when choosing a conference might include the following. If the SoTL work falls into a particular category, for example, service learning, use of technology in teaching, or investigating a particular type of pedagogy, there may be a conference that focuses on that topic. Some of these conferences, both discipline-specific and interdisciplinary, publish conference proceedings, so presenting at one of those could lead rather directly to a publication. Certain conferences, for example the Lilly Conferences on College and University Teaching, have a reputation for being friendly and welcoming to newcomers, and are organized in a way that provides many networking opportunities. Some conferences offer workshops for learning how to do SoTL.

Pause: Presenting a Paper—Questions to Consider

- What type of audience do I want for my presentation?
- Is there an on-campus venue to talk about my project, where I might get feedback before making a more formal conference presentation off-campus?
- Would I prefer a conference where I could get feedback from a wide range of disciplines?
- Would I prefer a conference where I might obtain mentoring for my work?
- Would I prefer a conference that will publish a proceedings?
- Am I able to get financial support for presenting at any type of conference or only conferences in my discipline?

Publishing an Article

STEM researchers will find that many of the comments we make about publishing SoTL articles are similar to comments that could be made about publishing in their disciplines. Writing and publishing an article requires more effort than making a conference presentation. It may also be more satisfying and will provide a permanent record of the work, one that can be a foundation for future improvements in teaching and learning by others. The likelihood of an article getting published can be greatly increased by the initial planning of the investigation. A well-designed study that is appropriately situated within the current literature on the question and has treated its subjects ethically provides a strong foundation for writing about the results.

The results and implications of some studies may be of interest to different audiences; these studies have the potential to generate more than one type of article. While most journals require that manuscripts be original work not published (or even submitted concurrently for consideration) elsewhere, authors may decide to write about different aspects of the study or its outcomes in more than one venue. For example, once a paper is accepted in a research journal, a less formal account of the implications of the research can be published elsewhere for practitioners or policy makers.

Choosing a Journal

For those beginning in SoTL, lack of familiarity with publishing education research can add to the difficulty of choosing a journal. Kennesaw State University's CETL provides a useful resource at https://cetl.kennesaw.edu/teaching-journals-directory that identifies journals that publish SoTL (on the pull-down menu for Topic in Higher Education, select "Pedagogical Research: General"). This website also lists disciplinary specific journals (including various STEM fields) and general or interdisciplinary journals on teaching, most with a link to the journal's website. It includes a number of international journals. For an extensive list of higher education journals, some with a STEM focus, located in the United Kingdom, Australia, Canada, and Europe, consult D'Andrea and Gosling (2005, Appendix A). Additional publishing options may result from an Internet search using the phrase: journals for publishing [insert STEM discipline] education.

We offer the following suggestions for factors to consider when choosing a journal. First, identify the target audience. Is it college professors in a particular

STEM discipline, researchers in STEM education, faculty members who prepare future STEM teachers, or a more general audience? Would faculty members in other disciplines be interested in the study? This may help decide between a journal focused on disciplinary education research and a journal that publishes SoTL for all disciplines. It is possible to send a query letter to a journal editor containing an abstract or short description of the article and ask for quick opinion of its appropriateness for that journal. Belcher (2009, 132–3) suggested including these additional elements in any query letter: why the editor and journal readers might be interested in the article, any grants or awards that supported the research, and a question aimed at determining the chance of rejection. For example, an author might write: *My paper reports on a situated study with only 25 subjects, whereas many of the articles in the journal involve larger sample sizes. Would this be a deterrent to its consideration?*

After identifying the target audience, there are many other factors that can be considered. Is it important that the article counts for tenure or promotion decision? Will a publication in a SoTL journal count? What journals were included in the literature review for the study? Are any journals from the literature review good publication venues for this article?

Once a potential journal is identified, new considerations arise. Does the journal publish research or practice or both? Will it accept case studies, or does it require experimental studies? Journals with a research focus may have a higher standard for evidence and require the work to be carefully situated within a theoretical framework. Visit the websites of potential journals and read the instructions or guidelines for authors, and if possible, some sample articles. Will the article meet the length limitations? Other factors to consider (or not) are acceptance rates, times to publication, and the reputation of the journal (as given by impact factor or other measures). The journal website may provide this information; if not, try an Internet search for a specific journal or type of journal (e.g., search for acceptance rate for *Journal of Chemical Education* or acceptance rates for chemistry education journals).

Whether or not there are publication fees (also called page charges or subventions) may be a matter of concern. Publishing is a costly enterprise, particularly when there are print versions of the journal. Someone—the author, the reader, libraries through their subscriptions, or a professional society by their sponsorship—has to provide financial support for publication. If authors or their institutions are required to pay publication fees, it is important to know that the journal has a meaningful peer-review process; otherwise, the publication represents a commercial agreement rather than an academic dissemination.

Fortunately, **open access**, meaning articles are freely available on the Internet, is increasing, and some high-quality journals now provide open access.

When choosing a journal, not only select the best fit, but also identify second- and third-choice journals in case the first-choice journal rejects the article. Szydlik (2015) described in some detail the process he used to select a journal for his SoTL work on the mathematical beliefs held by liberal arts students. Taber et al. (2014, 307–11) included an analysis of journals that publish different types of chemistry education research.

> **Pause: Choosing a Journal—Questions to Consider**
> - Have I identified the primary audience for my work?
> - Does my work have implications for teaching and learning beyond a single discipline?
> - Is my article a better fit for a journal that publishes research or practice?
> - Have I considered publishing a separate account for a different audience?
> - Do I need to take into account how my peers view the reputation of the particular journal or type of journal I have selected?
> - Am I concerned about time to publication?

Other Publication Venues

In addition to journals, there are other options for going public. A call for contributions to an edited volume may appear on a listserve, be announced at a conference, or come as a personal invitation from a colleague. There exist several well-regarded newsletters that are editor-reviewed, such as *The National Teaching and Learning Forum* and *The Teaching Professor*. Blogs about teaching provide a more informal option. The American Mathematical Society's blog, *On Teaching and Learning Mathematics* (http://blogs.ams.org/matheducation), is an example of an editor-reviewed blog. Submissions for potential posting to ISSOTL's blog (http://www.issotl.com/issotl15/blog) are reviewed by a subcommittee of the ISSOTL Communications Committee.

Getting the Manuscript Written and Submitted

We begin with general advice about writing that is derived from several sources as well as our own experience. We recommend attending to the results of Robert Boice (1990, 1992, 2000) who investigated which factors differentiated new faculty (including STEM faculty) who thrived from those who struggled in their first few years. In terms of writing articles for publication, he found that those who made a commitment to write in brief daily sessions (even as short as 15 minutes) were much more productive (as measured by pages written and articles published) than those who postponed writing until they had large blocks of uninterrupted time. Belcher (2009) and Rockquemore and Laszloffy (2008) also endorsed keeping to a regular writing schedule as key to success. These experts suggested finding a supportive community of writing peers. Silvia (2007) recommended starting a local writing group to provide accountability and research-related support and we have known colleagues for whom this approach was very helpful. Supportive writing communities can also be found on the Internet but the best advice may be from Rockquemore (2010), who suggested that each of us ask ourselves "what kind of writing support do I need?"

When preparing the manuscript use the format and adopt the style of writing found in articles that appear in the selected journal and follow instructions for submissions carefully. When writing for a SoTL journal, it is important to describe the problem or question that prompted the investigation clearly, provide context for the study relative to the current literature, describe how and from whom the evidence was gathered and how it was analyzed or interpreted, and state and justify the conclusions drawn. If the study involves a *What works?* question, make comparisons by providing baseline data or data from a control group and acknowledge other solutions or interventions that may exist. There are usually discussions of the limitations of the study, implications for practice, and possible next steps for extending the research. Much of this would be considered standard practice for publishing STEM research. One difference for many STEM faculty members arises from students being the subjects of SoTL research: journals may require proof of IRB approval (discussed in Chapter 3, Human Subjects Considerations) of a study before they will publish an article containing student data.

Before submitting the manuscript, recruit some colleagues to read and comment on it and consider making revisions based on their comments. Should the paper be rejected by the first-choice journal, it can be resubmitted to another journal. Before resubmitting elsewhere, it would be wise to consider whether some revising is in order based on the reasons given for the rejection.

> **Pause: Writing and Submitting the Manuscript—Questions to Consider**
> - Do I have a writing schedule that works for me?
> - Would joining a writing group help me be a more productive writer?
> - Have I carefully followed the guidelines for submissions?
> - Have I asked at least one colleague to read and comment on my article?

Responding to Reviews or Rejection

Journal acceptance rates vary widely, but it is rare in SoTL for an article to be accepted for publication exactly as it was submitted. Sometimes rejections come quickly and directly from the editor who judges the article not to be a good fit for the journal. Other times the reviewers find flaws they consider significant. Another possibility is that the reviewers recommend that the author revise and resubmit. In all these cases some amount of feedback and rationale will be given for the decision, either by the editor or the reviewers, or both. Regardless of the decision, a negative emotional response is the natural human reaction to critical comments. However, peer review plays a key role in publishing SoTL research and the revision of the paper is an important part of the publication process. Not responding appropriately to the reviewers' comments could lead to a subsequent rejection of the paper. Even in the case of an outright rejection, it is worthwhile to consider the reviewers' and editors' comments and make appropriate revisions before submitting the article to another journal.

In order to increase the likelihood of publication, we offer these suggestions for responding to reviewers' comments and requests for revisions. After the initial reading and reaction to the comments, the author should set them aside for a day or so. It is important to remember that reviewers' comments are intended to improve the quality of the final version of the paper. Whether or not a suggested change will improve the paper, as long as it does no detriment, the best course is to make the requested change. Each comment should receive a respectful response. The easiest way to do this is for the author to list each reviewer comment, followed by his or her response. It could be an agreement to make the requested change or revision, along with specific details of what was changed and where it occurred. For example, if the reviewer says: "how the evidence was gathered is not clear," rather than responding "we clarified how the evidence was gathered," it is much better to say: "we added more detail about how the evidence was gathered on page 14, lines 8–22." That makes it clear to the editor and the reviewer what was changed, and they can easily find the new or changed text if they want to examine it. Disagreement with a reviewer is also an option, but the author should explain, in a polite and factual manner, why the change was not made. Be sure to note and adhere to any deadlines that are given for response and revision.

> **Pause: Responding to Reviews or Rejection—Questions to Consider**
>
> - Have I responded to each comment that was made?
> - Do my responses make it easy for the reviewer and editor to find my revisions?
> - Do I explain my reasons for refusing to make certain changes?
> - Is my response respectful?
> - Have I noted and met any deadlines for revision?

Advice from Journal Editors

We close this chapter by letting several journal referees and an editor have the last word. Milton Cox (2013), founder and editor in chief of the *Journal on Excellence in College Teaching*, recorded a 20-minute program in which he discussed common reasons for manuscripts failing to get published, even though the projects looked promising. These included *not*:

- Defining the problem clearly and explaining why it was a problem
- Establishing a baseline at the beginning of the project
- Including a literature search and in it indicating how any intervention is connected to but different from what others have done
- Providing assessment results
- Acknowledging other solutions or approaches are possible.

For its fifth anniversary issue, the editor of the *International Journal for the Scholarship of Teaching and Learning* (IJ-SoTL) asked several referees to write 500 words on what characterizes an exemplary IJ-SoTL article (Maurer 2011; Rogers 2011; Simmons 2011; Stefani 2011; Tagg 2011). Four of the referees mentioned both the importance of a literature search and the need for clear, focused, jargon-free writing. Three commented how the interdisciplinary and international nature of the journal influenced their reviewing—specifically, that studies and results should apply or at least be interpretable beyond a single discipline. Authors considering submitting their work to an interdisciplinary journal would be well-served to examine the advice given by these IJ-SoTL referees.

REFERENCES

Belcher, W. 2009. *Writing Your Journal Article in Twelve Weeks: A Guide to Academic Publishing Success*. Thousands Oaks: SAGE Publications.

Bennett, C., and J. Dewar. 2004. "Carnegie Chronicle: What You Really Ought to Know about Collaboration on a SoTL Project." *National Teaching and Learning Forum 14* (1): pp. 4–6.

Boice, R. 1990. *Professors as Writers: A Self-Help Guide to Productive Writing*. Stillwater: New Forums Press.

Boice, R. 1992. *The New Faculty Member: Supporting and Fostering Professional Development*. San Francisco: Jossey-Bass.

Boice, R. 2000. *Advice to New Faculty: Nihil Nimus (Nothing in Excess)*. Boston: Allyn and Bacon.

Cox, M. 2013. *How Do I Prepare a SoTL Article for Publication?* Madison: Magna Publications.

D'Andrea, V., and D. Gosling. 2005. *Improving Teaching and Learning in Higher Education: A Whole Institution Approach*. Maidenhead: Open University Press.

Dewar, J., and C. Bennett, editors. 2015. *Doing the Scholarship of Teaching and Learning in Mathematics*. Washington: Mathematical Association of America.

Higgs, B. 2009. "The Carnegie Catalyst: A Case Study in the Internationalization of SoTL." *International Journal for the Scholarship of Teaching and Learning 3* (2), Article 4. doi: 10.20429/ijsotl.2009.030204.

Maurer, T. 2011. "Reviewer Essay: On Publishing SoTL Articles." *International Journal for the Scholarship of Teaching and Learning 5* (1), Article 32. doi: 10.20429/ijsotl.2011.050132.

Rockquemore, K. A. 2010. "Shut Up and Write!" *Inside Higher Ed*. June 14. https://www.insidehighered.com/advice/2010/06/14/shut-and-write.

Rockquemore, K. A., and T. Laszloffy. 2008. *The Black Academic's Guide to Winning Tenure—Without Losing Your Soul*. Boulder: Lynne Rienner Publishers.

Rogers, P. 2011. "Reviewer Essay: What Makes a Great Article for IJ-SoTL." *International Journal for the Scholarship of Teaching and Learning 5* (1), Article 33. doi: 10.20429/ijsotl.2011.050133.

Sawrey, B. 2008. "Collaborative Projects: Being the Chemical Education Resource." In *Tools of Chemistry Education Research*, edited by D. Bunce and R. Cole, pp. 201–14. Washington: American Chemical Society.

Silvia, P. 2007. *How to Write a Lot: A Practical Guide to Productive Academic Writing*. Washington: American Psychological Association.

Simmons, N. 2011. "Reviewer Essay: Exemplary Dissemination: Sowing Seed in IJ-SoTL." *International Journal for the Scholarship of Teaching and Learning* 5 (1), Article 35. doi: 10.20429/ijsotl.2011.050135.

Stefani, L. 2011. "Reviewer Essay: What Makes for a High Quality IJ-SoTL Research Article?" *International Journal for the Scholarship of Teaching and Learning* 5 (1), Article 36. doi: 10.20429/ijsotl.2011.050136.

Szydlik, S. 2015. "Liberal Arts Mathematics Students' Beliefs about the Nature of Mathematics: A Case Study in Survey Research." In *Doing the Scholarship of Teaching and Learning in Mathematics*, edited by J. Dewar and C. Bennett, pp. 145–65. Washington: Mathematical Association of America.

Taber, K., M. Towns, and D. Treagust. 2014. "Preparing Chemistry Education Research Manuscripts for Publication." In *Tools of Chemistry Education Research*, edited by D. Bunce and R. Cole, pp. 299–334. Washington: American Chemical Society.

Tagg, J. 2011. "Reviewer Essay: What Makes for a High Quality SoTL Research Article?" *International Journal for the Scholarship of Teaching and Learning* 5 (1), Article 34. doi: 10.20429/ijsotl.2011.050134.

Tsang, A. 2010. "Pitfalls to Avoid in Establishing A SoTL Academic Pathway: An Early Career Perspective." *International Journal for the Scholarship of Teaching and Learning* 4 (2), Article 19. doi: 10.20429/ijsotl.2010.040219.

CHAPTER 9

Reflecting on the Benefits of the Scholarship of Teaching and Learning

Twenty-five years after the publication of *Scholarship Reconsidered*, the Carnegie Foundation for the Advancement of Teaching republished the book in an expanded edition (Boyer 2015). In addition to the original text, it contains new chapters by nine higher education scholars who explored the impact of *Scholarship Reconsidered* (Boyer 1990) on faculty development, doctoral and professional education, rank and tenure systems, and academic disciplines. They concurred that "*Scholarship Reconsidered* raised our collective consciousness as to the nature and purpose of an academic vocation" (161). Moreover, they found Boyer's central question, "how the priorities of the professoriate can be better aligned with the full range of missions of higher education" (*xxiii*) equally pressing today in light of the changing landscape of higher education. The addition of more non-tenure track and adjunct positions, some of which carry expectations for scholarly work, has significantly shifted the profile of the academic workforce (Kezar and Maxey 2013). Student bodies have become more diverse, with ever-larger percentages of secondary students entering college (KewalRamani et al. 2007). Technology has offered many new options for instruction. Neuroscience has made discoveries about the biological basis of learning (Bransford et al. 2000; Doyle and Zakrajsek 2013; Leamnson 1999; Zull 2002). Calls have been made for the use of **high impact practices** (e.g., community-based learning, writing intensive courses, undergraduate

---

The Scholarship of Teaching and Learning, Jacqueline M. Dewar, Curtis D. Bennett, and Matthew A. Fisher.
© Oxford University Press, 2018. Published 2018 by Oxford University Press

research) to promote more engaged learning (Brownell and Swaner 2010; Kuh 2008). Each of these developments (as well as unforeseen future developments) has implications for teaching that open avenues for scholarship of teaching and learning (SoTL) investigations and for improving practice.

SoTL has something to offer faculty and their institutions as they wrestle with these developments. Participation in SoTL promotes more reflective teaching and improved teaching effectiveness. It is common for the details of what was good and bad in student work from previous semesters to fade from an instructor's memory. But, reflecting on student work from a SoTL perspective helps instructors capture the details. By asking and answering SoTL questions, faculty members can find out how well they are teaching and their students are learning, and they can gain insights for making improvements in their classrooms. As a consequence, SoTL offers a means other than student or peer evaluations to document teaching effectiveness and student learning in applications for merit, tenure, or promotion. On a more personal level, SoTL investigations can be energizing and deeply rewarding to faculty members at all stages of their careers, in part because of the collegial connections SoTL fosters across disciplines and institutions.

SoTL's impact on classroom teaching, professional development, institutional assessment, and the recognition and reward of pedagogical work has been documented (Burns 2017; Hutchings et al. 2011). SoTL can lead to institutional involvement in national or even international higher education initiatives as described in Gurm and Valleau (2010). Publicly embracing SoTL is one way for an institution to demonstrate its student-centeredness. In addition, because faculty members who ask and attempt to answer SoTL questions have to gather and analyze evidence that goes beyond grades on assignments and tests, they can help drive institutional assessment efforts to be a more meaningful process aimed at curriculum development and pedagogical improvement. In particular, when discussing curricular issues in a department, SoTL work can be a valuable ally. All in all, it seems that SoTL provides a method for addressing a number of the challenges facing departments, institutions, and the higher education community at large.

We have just presented general illustrations of the benefits of SoTL. In the domain of mathematics, 30 researchers involved in 15 SoTL projects reported how their work informed or benefitted their teaching, their careers, their departments, or other institutional efforts (Dewar and Bennett 2015). A synthesis of their reports led to a description of three **pathologies of *teaching*—amnesia, fantasia,** and **inertia**—and showed how SoTL can operate as an antidote to these (Bennett and Dewar 2015). The labels were borrowed from Shulman's (1999) **taxonomy for pitfalls of** *student learning*, where amnesia referred to students forgetting what they had learned, fantasia to persistent misconceptions, and inertia to the inability to use what was learned. In the pathologies of *teaching*, amnesia refers to how classroom insights fade from semester to semester (similar to Shulman's **pedagogical amnesia** as cited in Hutchings 1998, 17). Fantasia indicates how professors (and institutions) often have inaccurate beliefs regarding student thinking, behavior, or even what is happening in the classroom, while inertia signifies how difficult it can be for faculty to alter teaching methods when overwhelmed by the other demands of their jobs. These pathologies apply across all disciplines.

Each type of question in the SoTL taxonomy fits rather nicely as an antidote for one of the pathologies and can move us toward the ultimate goal of improving student learning. *What works?* questions are a natural way to combat pedagogical amnesia. Such questions push us to document and analyze a class (or program) in a more permanent and public way. The ability to revisit evidence and analyses months and even years later provides us with continued (and potentially even deeper) insights. For example, one author (Bennett 2016) revisited his first SoTL project after 15 years had passed and discovered flaws in his memory of what had happened in the class. Moreover, by going back to the rich information gathered during the project, he gained new insights regarding student behaviors.

What is? questions directly address pedagogical fantasia. By requiring us to find evidence and analyze what is happening, we often discover our mistaken beliefs. Uri Treisman's (1992) study (discussed in Chapter 2, Identifying Assumptions) is a perfect example of how a careful study of a *What is?* question led to dramatic changes in beliefs about students. Few studies will make as dramatic a change in our understandings, but each time we identify a misconception, we gain understanding and improve our ability to teach our students.

What could be? questions prompt us to break free of past patterns, identify new conceptualizations of courses, and experiment. John Holcomb (2015) asked precisely such a question to create and study changes in his elementary statistics course. He discussed how his SoTL project motivated him to design new components for his class and change his assessment and teaching.

Despite these natural linkings, question types do cross the boundaries. *What is?* and *What works?* questions can shift us out of old patterns of teaching. *What could be?* and *What is?* questions lead us to analyze and keep gathering data to fight amnesia. And *What works?* and *What could be?* questions may lead to discoveries of misconceptions. A final reason to value SoTL is for its ability to turn the work of teaching into the joy of discovery.

REFERENCES

Bennett, C. 2016. "Effects of a Capstone Course on Future Teachers (and the Instructor): How a SoTL Project Changed a Career." In *Mathematics Education: A Spectrum of Work in Mathematical Sciences Departments*, edited by J. Dewar, P.-s. Hsu, and H. Pollatsek, pp. 43–54. New York: Springer.

Bennett, C., and J. Dewar. 2015. "Synthesis of the Value and Benefits of SoTL Experienced by the Contributors." In *Doing the Scholarship of Teaching and Learning in Mathematics,* edited by J. Dewar and C. Bennett, pp. 203–6. Washington: Mathematical Association of America.

Boyer, E. L. 1990. *Scholarship Reconsidered: Priorities of the Professoriate*. San Francisco: Jossey-Bass.

Boyer, E. L. 2015. *Scholarship Reconsidered: Priorities of the Professoriate*, expanded ed. San Francisco: Jossey-Bass.

Bransford, J., A. Brown, and J. Pellegrino, editors. 2000. *How People Learn: Brain, Mind, Experience, and School*. Washington: National Academy Press.

Brownell, J., and L. Swaner. 2010. *Five High-Impact Practices: Research on Learning Outcomes, Completion, and Quality*. Washington: Association of American Colleges and Universities.

Burns, K. 2017. "Community College Faculty as Pedagogical Innovators: How the Scholarship of Teaching and Learning (SoTL) Stimulates Innovation in the Classroom." *Community College Journal of Research and Practice* 41 (3): pp. 153–67. doi: 10.1080/10668926.2016.1168327.

Dewar, J., and C. Bennett. 2015. *Doing the Scholarship of Teaching and Learning in Mathematics*. Washington: Mathematical Association of America.

Doyle, T., and T. Zakrajsek. 2013. *The New Science of Learning: How to Learn in Harmony with Your Brain*. Sterling: Stylus.

Gurm, B., and A. Valleau. editors. 2010. "The Attraction, Value and Future of SoTL: Perspectives from the Carnegie Affiliates." [Special Issue] *Transformative Dialogues: Teaching and Learning eJournal* 4 (1). Accessed August 31, 2017. http://www.kpu.ca/td/past-issues/4-1.

Holcomb, J. 2015. "Presenting Evidence for the Field that Invented the Randomized Clinical Trial." In *Doing the Scholarship of Teaching and Learning in Mathematics,* edited by J. Dewar and C. Bennett, pp. 117–26. Washington: Mathematical Association of America.

Hutchings, P. 1998. "Defining Features and Significant Functions of the Course Portfolio." In *The Course Portfolio: How Faculty Can Examine Their Teaching to Advance Practice and Improve Student Learning,* edited by P. Hutchings, pp. 13–18. Washington: AAHE.

Hutchings, P., M. Huber, and A. Ciccone. 2011. *Scholarship of Teaching and Learning Reconsidered*. San Francisco: Jossey-Bass.

Leamnson, R. 1999. *Thinking about Teaching and Learning: Developing Habits of Learning with First Year College and University Students*. Sterling: Stylus.

KewalRamani, A., L. Gilbertson, M. Fox, and S. Provasnik. 2007. *Status and Trends in the Education of Racial and Ethnic Minorities, NCES 2007-039*. Washington: National Center for Education Statistics, Institute of Education Sciences, U.S. Department of Education.

Kezar, A., and D. Maxey. 2013. "The Changing Academic Workforce." *Trusteeship Magazine 21* (3). Accessed December 11, 2017. https://www.agb.org/trusteeship/2013/5/changing-academic-workforce.

Kuh, G. 2008. *High-Impact Educational Practices: What They Are, Who Has Access to Them, and Why They Matter*. Washington: Association of American Colleges and Universities.

Shulman, L. 1999. "Taking Learning Seriously." *Change 31* (4): pp. 10–17.

Treisman, U. 1992. "Studying Students Studying Calculus: A Look at the Lives of Minority Mathematics Students in College." *College Mathematics Journal 23* (5): pp. 362–73.

Zull, J. E. 2002. *The Art of Changing the Brain: Enriching the Practice of Teaching by Exploring the Biology of Learning*. Sterling: Stylus Publishing.

APPENDIX I

Carnegie CASTL Scholars in STEM and Related Fields

We have included scholars from biology, chemistry/biochemistry, civil engineering, computer science, earth science, electrical engineering, geology, mathematics, mechanical engineering, medical and allied health fields, physics, physics education, science education, and statistics. Institutional affiliations shown below are those at the time of their CASTL (Carnegie Academy for the Scholarship of Teaching and Learning) scholar year.

1998–1999 Scholars

Brian Coppola
Department of Chemistry
University of Michigan
Ann Arbor, MI

James Hovick
Department of Chemistry
University of North Carolina at Charlotte
Charlotte, NC

Deborah Wiegand
Department of Chemistry
University of Washington
Seattle, WA

1999–2000 Scholars

Peter Alexander
Mathematics
Heritage College
Toppenish, WA

Thomas Banchoff
Mathematics Department,
Brown University
Providence, RI

Bruce Cooperstein
Department of Mathematics
University of California, Santa Cruz
Santa Cruz, CA

Linda Hodges
Chemistry Department
Agnes Scott College
Decatur, GA

Dennis C. Jacobs
Department of Chemistry and Biochemistry
University of Notre Dame
Notre Dame, IN

Anita Salem
Department of Mathematics
Rockhurst University
Kansas City, MO

Mark Walter
Department of Chemistry
Oakton Community College
Des Plaines, IL

2000–2001 Scholars

Curtis Bennett
Department of Mathematics and Statistics
Bowling Green State University
Bowling Green, OH

Jack Bookman
Department of Mathematics
Duke University
Durham, NC

Hessel Bouma III
Biology Department
Calvin College
Grand Rapids, MI

Mary Burman
School of Nursing
University of Wyoming
Laramie, WY

William Cliff
Department of Biology
Niagara University
Niagara University, NY

JoLaine Draugalis
College of Pharmacy
University of Arizona
Tucson, AZ

Maura Flannery
Department of Computer Science, Mathematics, and Science
St. John's University
Jamaica, NY

John Holcomb, Jr
Department of Mathematics
Cleveland State University
Cleveland, OH

Craig Nelson
Department of Biology
Indiana University
Bloomington, IN

Marilyn Repsher
Department of Mathematics
Jacksonville University
Jacksonville, FL

David Takacs
Department of Earth Systems Science and Policy
California State University, Monterey Bay
Marina, CA

Emily van Zee
Science Education, Department of Curriculum and Instruction
University of Maryland, College Park
College Park, MD

2001–2002 Scholars

Harvey Bender
Department of Biological Sciences
University of Notre Dame
Notre Dame, IN

Spencer Benson
Department of Cell Biology and Molecular Genetics
University of Maryland College Park
College Park, MD

Alix Darden
Biology Department
The Citadel
Charleston, SC

Amy Haddad
School of Pharmacy and Allied Health Professions
Creighton University
Omaha, NE

Mangala Joshua
Physical Science Department
Mesa Community College
Mesa, AZ

Patrick Kenealy
Department of Physics/Science Education
California State University, Long Beach
Long Beach, CA

Anthony Marchese
Department of Mechanical Engineering,
Rowan University
Glassboro, NJ

Charles McDowell
Department of Computer Science
University of California, Santa Cruz
Santa Cruz, CA

Donald Misch
Associate Professor
Department of Psychiatry and Health Behavior
Medical College of Georgia
Augusta, GA

Melinda Piket-May
Electrical and Computer Engineering Department
University of Colorado at Boulder
Boulder, CO

Steven Pollock
Department of Physics
University of Colorado at Boulder
Boulder, CO

2003–2004 Scholars

Michael Axtell
Department of Mathematics and Computer Science
Wabash College
Crawfordsville, IN

Curtis Bennett
Department of Mathematics
Loyola Marymount University
Los Angeles, CA

S. Raj Chaudhury
BEST Lab/Physics Department
Norfolk State University
Norfolk, VA

Jacqueline Dewar
Department of Mathematics
Loyola Marymount University
Los Angeles, CA

Heidi Elmendorf
Biology Department
Georgetown University
Washington, DC

Michael Loui
Department of Electrical and Computer Engineering
University of Illinois at Urbana-Champaign
Urbana, IL

Whitney Schlegel
School of Medicine
Indiana University Bloomington
Bloomington, IN

Kathy Takayama
School of Biotechnology and Biomolecular Sciences, Microbiology and Immunology
The University of New South Wales
Sydney, Australia

2005–2006 Scholars

Michael Burke
Department of Mathematics
College of San Mateo
San Mateo, CA

Tricia Ferrett
Department of Chemistry
Carleton College
Northfield, MN

Matthew Fisher
Department of Chemistry
Saint Vincent College
Latrobe, PA

Lorraine Fleming
Department of Civil Engineering
Howard University
Washington, DC

Bettie Higgs
Department of Geology
University College Cork
Cork, Ireland

Gregory Kremer
Department of Mechanical Engineering
Ohio University
Athens, OH

Crima Pogge
Department of Biology
City College of San Francisco
San Francisco, CA

Joanne Stewart
Department of Chemistry
Hope College
Holland, MI

APPENDIX II

Participants in the Carnegie Scholarly and Professional Societies Program

Carnegie Scholarly and Professional Societies Program Participants:
- American Academy of Religion
- Association of American Geographers
- American Association of Colleges of Nursing (AACN)
- American Association of Colleges of Pharmacy
- American Association of Philosophy Teachers
- American Dental Education Association (ADEA)
- American Economic Association (AEA)
- American Historical Association
- American Institute of Biological Sciences
- American Philosophical Association
- American Physical Society (APS)
- American Physical Therapy Association (Education Section)
- American Political Science Association
- American Psychological Association
- American Society for Microbiology
- American Sociological Association
- Association for Gerontology in Higher Education
- Association for Integrative Studies
- Association for Theatre in Higher Education
- Association of American Law Schools

- Centre for Health Sciences and Practice
- College Music Society
- Conference on College Composition and Communication
- Conference on English Education
- Mathematical Association of America
- National Communication Association
- National Council of Teachers of English/College Section
- Two-Year College English Association of NCTE

APPENDIX III

Sample Focus Group Protocol

This example is derived from a focus group that was held at the end of the semester to assess the effect of introducing group projects on community-based issues into MATH 102, a quantitative literacy course (Dewar et al. 2011). The example includes: (1) a list of items to gather or prepare in advance, (2) advice on set-up, (3) a timeline of activities for the focus group, (4) questions for a paper survey to be administered to early arrivals, (5) a form for the roundtable ranking activity, (6) text for PowerPoint slides used to conduct the focus group, and (7) the protocol with a script for the facilitator conducting the focus group.

1. **Items to prepare or gather in advance**

   - Room with large table or movable chairs and desks
   - Informed consent forms, if not obtained in advance when recruiting subjects
   - Sign-in sheet (to confirm that all participants gave informed consent)
   - Paper surveys for early arrivals, if used (one per participant)
   - Pencils, one for each participant
   - Projector and computer with prepared PowerPoint slides (see text for slides in 6. Text for PowerPoint slides)
   - Numbered "participant-signs," one for each participant (a tent card made from triple-folded 8.5"×11" paper, pre-numbered with participant numbers from #1 to #n, where n is the maximum number of participants)
   - 3"×5" index cards, at least one per participant for each planned index card rating activity

- Roundtable ranking activity forms, one for each group of three or four participants
- Recording device, if using, or note-taking materials for assistants

2. **Set-up** (Complete ten minutes before first participants are expected to arrive)

   - Arrange chairs or desks in a circle or horseshoe shape to accommodate planned number of attendees
   - Projector and computer
   - Numbered "participant-signs" (see description of participant-signs in 1. Items to prepare or gather in advance) placed at each seat
   - Index cards at each place (at least one per participant for each planned index card rating activity)
   - Pencil or pen at each place
   - Food (if providing)
   - Roundtable activity sheets off to the side
   - Write schedule on the board (or post on wall on large poster)

3. **Timeline of activities**

   12:15–12:30 Survey and pizza
   12:30 Introduction and index card rating activity #1
   12:40 Roundtable activity generating and ranking strengths and weaknesses
   12:50 Index card rating activity #2
   1:00–1:15 Open-ended questions (round-robin format)
   1:15 End and thank you

4. **Paper survey questions for early arrivals**

Respond to each statement by circling your level of agreement or disagreement:

SA = Strongly Agree   A = Agree   N = Neutral   D = Disagree   SD = Strongly Disagree

After taking MATH 102 I am more aware of the usefulness of mathematics in addressing real problems.   SA A N D SD
Explain briefly:

After taking MATH 102 I am more aware of community or environmental issues on campus or near campus.   SA A N D SD
Explain briefly:

The MATH 102 project helped me connect my classroom learning to the real world.   SA A N D SD
Explain briefly:

The MATH 102 project helped me practice and learn mathematical or analytical skills.   SA A N D SD

Explain (or give examples if you can):

The MATH 102 project taught me non-mathematical skills.   SA A N D SD

Explain (or give examples if you can):

Two short-answer open response questions:

If next semester the math faculty were going to change everything about the MATH 102 course, except ONE THING, what ONE THING would you recommend they KEEP THE SAME:

If next semester the math faculty were going to keep everything about MATH 102 course the same, except ONE THING, what ONE THING would you recommend they CHANGE.

5. Roundtable ranking activity form

The form is an 8.5"× 11" sheet of a paper with the following information on the two sides (Side A and Side B).

Side A:

List the assigned participant numbers of the people in the group: #_____

Roundtable Ranking of Strengths (SIDE A)

Passing this sheet of paper rapidly from person to person in the group, each person writes a **strength** of this course, saying it aloud while writing. If you can't come up with a new strength, you may pass the paper on without writing, but try to keep going until time is up.

(Leave about a half-page of open space here.)

Working as a team, please rank order the strengths you identified with the most important ones at the top of your list. Please give at least three:

#1
#2
#3
#4
etc.

Side B:

Roundtable Ranking of Weaknesses (SIDE B)

Passing this sheet of paper rapidly from person to person in the group, each person writes a **weakness** of this course, something that might be changed or improved, saying it aloud while writing. If you can't come up with a new weakness, you may pass the paper on without writing, but try to keep going until time is up.

(Leave about a half-page of open space here.)

Working as a team, please rank order the drawbacks of the course—things you would change—with the most critical ones at the top of your list. Please give at least three:

#1
#2
#3
#4
etc.

6. Text for PowerPoint slides

Title Slide

Focus Group on MATH 102
Names of Focus Group Facilitator and Assistant(s), if any
Date

Slide #1

This Focus Group is

- **Anonymous:** None of your comments will be attributed to you individually. Please use the assigned number when making comments and referring to other students.
- **Candid:** We want honest, constructive information.
- **Confidential:** We will report only to appropriate parties. Please do not repeat any participant comments outside of this room.
- **Productive:** Faculty will consider the information you provide when deciding on changes.

Slide #2

Index card rating #1:
Put your assigned participant number (from your tent card) in upper right-hand corner.

Slide #3

On the index card:
Write one word or phrase to describe your impressions of the MATH 102 course.

Below this word, please **write a number** from 1 to 5 that describes your satisfaction with the course:

1= Low 5 = High.

Slide #4

Roundtable: Generating Strengths (Side A).

Passing this sheet of paper, rapidly from one person to the next, on Side A, each person writes down his or her participant number followed by a strength of the MATH 102 course, saying it aloud while writing. You may pass without writing, but keep the paper going around.

Slide #5

Roundtable: Ranking 1–2–3.

As a team, at the bottom of Side A, please rank order the top three strengths you identified, with the most important strength at the top of your list.

Slide #6

Roundtable: Generating Weaknesses (Side B).

Passing this sheet of paper rapidly from one person to the next, on Side B, each person writes down his or her participant number followed by something about the course that might be changed or improved, saying it aloud while writing. You may pass without writing, but keep the paper going around.

Slide #7

Ranking 1–2–3:

As a team, at the bottom of Side B, please rank order the top three weaknesses—things you would change—with the most important one to change at the top of your list.

Slide #8

Index card rating #2:

Put your participant number in upper right-hand corner, and under that write your project topic. (Note: A bulleted list of the project topics from the course was provided.)

Slide #9

On the index card:

Write one word or phrase to describe your impressions of the project you worked on.

Below this word, please **write a number** from 1 to 5 that describes your satisfaction with the project as a teaching/learning tool:
1= Low 5 = High.

Slide #10

Open-ended Question (round-robin):
Suppose one of your friends (not in MATH 102) said:
Lots of what we have to study in college, math for instance, will never be used in the future.
How would you respond?

Slide #11

The End!

- Thank you so much!
- Your instructors will hear your comments anonymously (after they turn in their grades) and will consider them when revising the course (or whatever is the purpose for the focus group).

7. Protocol with facilitator's script for conducting the focus group

Before designated start time (12:30 p.m.)

- Room is prepared
- Assistant meets students as they arrive, checks off students' names, and gives out the paper surveys (and pencil if needed)
- Participants are admitted to the room as they arrive (up to a pre-set maximum number)
- Promptly, at 12:30, door closes, no one else is admitted, and focus group facilitator starts the focus group.

12:30 Introduction and index card rating activity #1: (Use the Title Slide and Slides #1–#3.)

[Words in italic are said by the focus group facilitator.]

Thank you for agreeing to participate in this focus group. I will be leading the focus group, helped by two assistants. We are interested in your opinions about the MATH 102 course you just completed this semester. This focus group is:

- **Anonymous:** *None of your comments will be attributed to you individually. Please use the assigned number on the tent card in front of each of you when making comments and referring to other students.*

- **Candid:** We want honest, constructive information.
- **Confidential:** We will report only to appropriate parties. Please do not share other participants' comments outside of this room.
- **Productive:** Faculty will consider the information you provide when deciding on changes.

We begin with the first index card rating activity:
Pick up one of the index cards near you and put your assigned number (shown on the tent card in front of you) in upper right-hand corner. Write one word or phrase to describe your impressions of the MATH 102 course. Below this word, please write a number from 1 to 5 that describes your satisfaction with the course: 1= Low and 5 = High.

After it appears (almost) everyone has stopped writing, ask participants to report out by saying: *Let's hear some of your responses and tell us a little about why you said the phrase you did and rated the way you did. Remember to state your number before you speak. Who would like to start?*

After the first person speaks, say *Thank you* and ask, *Is there another volunteer?* Assistant(s), if used, take notes.

Signal the opportunity to speak on this topic is coming to an end by saying, *Let's hear from two more.*

12:40 Roundtable activity generating and ranking strengths and weaknesses: (Use Slides #4–#7.)

Divide participants into groups of four (with one or two groups of three, if needed, but do not have groups of just two) and distribute one roundtable ranking activity form to each group. Give these directions: *At the top of Side A, write down the number of each person in the group. Then, passing the form rapidly from one person to the next, each person should write down his or her participant number followed by one strength of the MATH 102 course, saying it aloud while writing it down. Pass the form to the left, so the next person can write a strength and say it aloud. If you can't think of a new strength, you may say "pass." Keep the form going around trying to generate more strengths until the time is up. Remember to write down your participant number in front of each strength you write down. You will have two minutes.* Let them continue for two minutes; then give the following instructions.

As a team, at the bottom of Side A, please rank order the top three strengths you identified, with the most important strength at the top of your list. Continue this for about two minutes, then have the groups repeat the exercise, using Side B to list and rank the weaknesses by giving the following directions.

Next, turn the form over to Side B. Then, passing the form rapidly from one person to the next, each person should write down his or her participant number followed by a weakness of the MATH 102 course, something that might be changed or improved, saying it aloud while writing it down. Pass the form to the left, so the next person can write a weakness and say it aloud. If you can't think of a new weakness, you may say "pass." Keep the form going around trying to generate more weaknesses until the time is up. Let them continue for two minutes; then give the following instructions.

As a team, at the bottom of Side B, please rank order the top three weaknesses you identified, with the most important one to change at the top of your list. Let them continue for two minutes, then collect the roundtable sheets. There is no report out on this activity.

12:50 Index card rating activity #2: (Use Slides #8 and #9)

We have another index card rating activity. Pick up one of the index cards near you and put your assigned number in upper right-hand corner; under that write the name of your project topic from the bulleted list on the slide.

Write one word or phrase to describe your impressions of the project you worked on. Below this word, please write a number from 1 to 5 that describes your satisfaction with the project as a teaching/learning tool: 1= Low and 5 = High.

Report out after it appears (almost) everyone has stopped writing: *Let's hear some of your responses and tell us a little about why you said the phrase you did and rated the way you did. Remember to state your number before you speak.*

Assistant(s), if used, take notes:

As before, signal the opportunity to speak on this topic is coming to an end by saying, *Let us hear from two more.*

1:00 Open-ended questions (round-robin format): (Use Slide #10)

Here is a question: Suppose one of your friends (not taking MATH 102) said, "Lots of what we have to study in college, math for instance, will never be used in the future." Take a moment to think how you would respond.

After 30 to 60 seconds give the following directions: *We will hear your responses in a round-robin format. That means we will hear from one volunteer, and then the opportunity to speak will move clockwise around the room. You do have the option to pass when it is your turn. May I have a volunteer to start? Remember to say your assigned number before speaking.*

Once everyone has had a chance to speak, ask: *Are there any additional comments you would like to make?*

Note: It is important to adhere to the schedule. So, while there may only be time for one open-ended question, it is a good idea to have several more prepared, just in case. The following are the additional questions we had ready (including having them on slides):

- What was the most important lesson you learned from your project?
- Would you advise faculty teaching MATH 102 in the future to keep the project component or to throw it out and put in one more mathematical topic instead? Explain why.
- Do you have any suggestions for:
  – improving the project timeline?
  – new project ideas?
  – assigning project groups or helping them work together better?
- Please identify the mathematical topic you learned about in MATH 102 that you consider the most useful or valuable. Explain why in a sentence or two.

1:15 End and thank you (Use Slide #11)

End on time by thanking the students and repeating these important points: *Thank you very much for your participation. Your instructors will hear your comments anonymously (after they turn in their grades) and will consider them when revising the course. And remember that you should not share participants' comments outside of this room.*

REFERENCES

Dewar, J., S. Larson, and T. Zachariah. 2011. "Group Projects and Civic Engagement in a Quantitative Literacy Course." *PRIMUS: Problems, Resources, and Issues in Mathematics Undergraduate Studies 21* (7): pp. 606–37.

Millis, B. 2004. "A Versatile Interactive Focus Group Protocol for Qualitative Assessment." In *To Improve the Academy*, Vol. 21, edited by C. M. Wehlburg and S. Chadwick-Blossey, pp. 125–41. San Francisco: Jossey Bass.

APPENDIX IV

Instruments, Taxonomies, and Models for SoTL Studies in STEM Fields

These are not intended as exhaustive lists, nor is every item listed available for free. They are offered as an indication of the variety of instruments, taxonomies, and models that are available for scholarship of teaching and learning (SoTL) research in science, technology, engineering, and mathematics (STEM) fields. Because Internet addresses change over time, we list only the names of various surveys and scales. The reader can obtain more information about any of these by using a web search engine. For each of the taxonomies and models, we provide references.

Instruments

Anchoring Concepts Content Map (General Chemistry)
Assessment Resource Tools for Improving Statistical Thinking
Attitudes Toward Mathematics Inventory
Concept Inventories:
  Astronomy and Space Science Concept Inventory
  Biology Concept Inventory
  Biological Experimental Design Concept Inventory
  Calculus Concept Inventory
  Central Dogma Concept Inventory (Biology)
  Circuits Concept Inventory

Chemistry Concept Inventory
Computer Engineering Concept Inventory
Digital Logic Concept Inventory
Dynamics Concept Inventory
Electromagnetics Concept Inventory
Electronics Concept Inventory
Enzyme-Substrate Interactions Concept Inventory
Flame Test Concept Inventory
Fluid Mechanics Concept Inventory
Force Concept Inventory (FCI) (the original concept inventory)
Genetics Concept Inventory
Geoscience Concept Inventory
Heat Transfer Concept Inventory
Homeostasis Concept Inventory (Biology)
Meiosis Concept Inventory (Biology)
Precalculus Concept Inventory
Signals and Systems Concept Inventory
Statistical Reasoning in Biology Concept Inventory
Strength of Materials Concept Inventory
Thermodynamics Concept Inventory
Waves Concept Inventory
Comprehensive Assessment of Outcomes (CAOS)
Diagnostic of Undergraduate Chemistry Knowledge (DUCK)
Faculty Survey of Student Engagement (FSSE)
Fennema-Sherman Mathematics Attitudes Scales
Field-tested Learning Assessment Guide (FLAG)
Indiana Mathematics Belief Scales
Maryland Physics Expectations Survey
Mathematics Anxiety Rating Scale (MARS)
National Survey of Student Engagement (NSSE)
Pittsburgh Engineering Attitudes Survey
Student Assessment of Learning Gains (SALG)
Student Attitudes Toward Statistics (SATS)
Student Attitudes Toward STEM (S-STEM)
Views About Sciences Survey (VASS)

See also the Appendix of Chemistry Diagnostic Assessments and Concept Inventories in Bretz (2014).

Taxonomies and Models

Bigg's Structure of the Observed Learning Outcome (SOLO) Taxonomy (Biggs and Collis 1982)
Bloom's Taxonomy (Original) (Bloom 1956)
Bloom's Taxonomy (Revised) (Anderson and Krathwohl 2001)
Dimensions of Learning Model (Marzano et al. 1993)
Integrated Problem-Solving Model (Litzinger et al. 2010)
Lifelong Learning Standards (Marzano et al. 1993)
Mathematical Knowledge-Expertise Taxonomy (Bennett and Dewar 2013)
Model of Epistemological Reflection (Baxter Magolda 1992)
Model of Domain Learning (Alexander 2003)
Model of Intellectual Functioning (Costa 1985)
Perry's Scheme of Intellectual and Ethical Development (Perry 1970)
Taxonomy of the Affective Domain (Krathwohl et al. 1964)
Taxonomy of Significant Learning Experiences (Fink 2003)
Taxonomy of SoTL Questions (Hutchings 2000)
Typology of Scientific Knowledge (Shavelson and Huang 2003)
Women's Ways of Knowing (Belenky et al. 1986)

REFERENCES

Alexander, P. 2003. "The Development of Expertise: The Journey from Acclimation to Proficiency." *Educational Researcher 32* (8): pp. 10–14.

Anderson, L., and D. Krathwohl. 2001. *Taxonomy for Learning, Teaching and Assessing: A Revision of Bloom's Taxonomy of Educational Objectives.* New York: Longman.

Baxter Magolda, M. 1992. *Knowing and Reasoning in College: Gender-Related Patterns in Students' Intellectual Development.* San Francisco: Jossey-Bass.

Belenky, M. F., B. M. Clinchy, N. R. Goldberger, and J. M. Tarule. 1986. *Women's Ways of Knowing.* New York: Basic Books.

Bennett, C., and J. Dewar. 2013. "SoTL and Interdisciplinary Encounters in the Study of Students' Understanding of Mathematical Proof." In *The Scholarship of Teaching and Learning In and Across the Disciplines,* edited by K. McKinney, pp. 54–73. Bloomington: Indiana University Press.

Biggs, J. B., and K. F. Collis. 1982. *Evaluating the Quality of Learning—the SOLO Taxonomy.* New York: Academic Press.

Bloom, B., editor. 1956. *Taxonomy of Educational Objectives: The Classification of Educational Goals, Handbook I: Cognitive Domain.* New York: McKay.

Bretz, S. 2014. "Designing Assessment Tools to Measure Students' Conceptual Knowledge of Chemistry." In *Tools of Chemistry Education Research*, edited by D. Bunce and R. Cole, pp. 156–68. Washington: American Chemical Society.

Costa, A. L. 1985. "Toward a Model of Human Intellectual Functioning." In *Developing Minds: A Resource Book for Teaching Thinking*, edited by A. L. Costa, pp. 62–5: Alexandria: Association for Supervision of Curriculum Development.

Fink, L. D. 2003. *Creating Significant Learning Experiences: An Integrated Approach to Designing College Courses*. San Francisco: Jossey-Bass.

Hutchings, P., editor. 2000. *Opening Lines: Approaches to the Scholarship of Teaching and Learning*. Menlo Park: The Carnegie Foundation for the Advancement of Teaching.

Krathwohl, D. R., B. S. Bloom, and B. B. Masia. 1964. T*axonomy of Educational Objectives: The Classification of Education Goals, Handbook II: The Affective Domain*. New York: McKay.

Litzinger, T., P. Van Meter, C. Firetto, L. Passmore, C. Masters, S. Turns, ..., S. Zappe. 2010. "A Cognitive Study of Problem Solving in Statics." *Journal of Engineering Education 99* (4): pp. 337–53. doi: 10.1002/j.2168-9830.2010.tb01067.x.

Marzano, R., D. Pickering, and J. McTighe. 1993. *Assessing Student Outcomes: Performance Assessment Using the Dimensions of Learning Model*. Alexandria: Association for Supervision and Curriculum Development. Accessed August 31, 2017. http://files.eric.ed.gov/fulltext/ED461665.pdf

Perry, W. G. Jr 1970. *Forms of Intellectual and Ethical Development in the College Years: A Scheme*. New York: Holt, Rinehart, and Winston.

Shavelson, R. J., and L. Huang. 2003. "Responding Responsibly to the Frenzy to Assess Learning in Higher Education." *Change 35* (1): pp. 10–19.

INDEX

accreditation
  engineering 10
affordances, *see* student affordances
alignment 41, 45, 46–7, 51, 63, 65, 66, 86, 103, 107, 115, 120, 122, 126
American Chemical Society 15, 135
American Physical Society 15, 135, 159
American Statistical Association 46
amnesia, *see* pathologies of teaching, amnesia
assessment
  alignment of 41, 46–7, 63, 65
  classroom assessment techniques 64, 65, 66
  institutional 26, 65, 69, 96, 115, 133, 146
  instruments 170–2
  of student learning (outcomes) 8, 10, 44, 45, 82, 101, 147
assignment design 45, 65–6
Association of American Colleges and Universities (AACU) 9, 115, 119

Basic and Applied Social Psychology 46
Berheide, Kate 2
bottlenecks 30, 32, 50
Boyer, Ernest 6–7, 9, 13, 15, 18, 145

Carnegie Academy for the Scholarship of Teaching and Learning (CASTL) 9

scholars 9, 151–8; *see also* Carnegie scholar(s)
Scholars Program ix, 9, 11, 23
Carnegie Foundation for the Advancement of Teaching 6, 7, 16, 40, 145
Carnegie Institutional Leadership and Affiliates Program 9, 105
Carnegie Institutional Leadership Clusters 9
Carnegie scholar(s) xiii, 2, 12, 44, 100, 132–3; *see also* Carnegie Academy for the Scholarship of Teaching and Learning (CASTL), scholars
Carnegie Scholarly and Professional Societies Program 9, 159
CASTL, *see* Carnegie Academy for the Scholarship of Teaching and Learning
classroom assessment techniques 64, 65–6
coding data 62, 64, 71, 90, 95, 112, 120–6, 127, 128
  a priori categories 121; *see also* coding data, predetermined categories
  codes 121, 123, 124, 125, 128
  common themes 71, 95, 112, 121
  constructed theories 127
  emergent categories 122–3, 124, 125, 126, 128; *see also* coding data, inductive categories
  focused 123
  inductive 122; *see also* coding data, emergent categories

initial 123, 124
inter-coder reliability 124, 125, 126, 127, 128
labels 121, 123; *see also* coding data, codes
memo-ing 123
predetermined categories 121–2, 124, 125, 126; *see also* coding data, a priori categories
Cohen's *d* 46, 49
community college(s) 2
concept map 99, 126
  coding of 126
confounding factors 42
content analysis, *see* coding data
control group 10, 31, 39, 40, 44, 49, 51, 127
  lack of 41–5
criterion-referenced evaluation 45
critical friend 133

data
  existing 54
  familiar sources of 62–4
  matched 48, 73
  qualitative 14, 60, 61, 64, 66, 68, 90, 91, 95, 99, 104, 111–12, 118, 123, 126, 128; *see also* standards for qualitative research
  quantitative 14, 60, 61, 62, 63, 66, 68, 75, 91, 104, 111, 112, 128; *see also* standards for quantitative research
  quantitative versus qualitative 61–2
  triangulating, *see* triangulation

INDEX

DBER, *see* discipline-based education research
Decoding the Disciplines 32, 50
discipline-based education research (DBER) 1, 5, 10–15, 17, 23
disciplines
  structure of 13, 122
dissemination
  choosing a journal 137–9
  conference presentation 135–6
  publication fees 137
  publication venues 139
  responding to reviews 141–2
  writing a manuscript 139–41
  see also going public

effect size 46, 48, 49; *see also* Cohen's *d*
absolute 48
engagement
  civic 47, 77, 133
  interactive 8, 48
  scholarship of, *see* scholarship, of engagement
  social 119
  student 66, 73, 171
Escalante, Jaime 8
evidence, *see* data
exempt study 53
expedited review 53
experimental (treatment) group 10, 31, 39, 40, 41, 42, 44, 49, 51, 127
experimental study 40, 49, 51, 127, 138

factor analysis 111
fantasia, *see* pathologies of teaching, fantasia
Felder, Richard 1
focus group 41, 86, 90, 92–6, 118
  index card activity 93–4, 166–7, 168
  model for conducting a 92–5
  roundtable ranking activity 94, 167–8
  sample protocol 161–9
  used in STEM education studies 90, 92–6, 118
Freudenthal, Hans 1

Freudenthal Institute for Science and Mathematics Education 1

going public 7, 8, 132, 134–43; *see also* dissemination
gold standard 40, 127; *see also* random assignment

high impact practices 145
HREB, *see* Human Research Ethics Board
HSRB, *see* Human Subjects Review Board
Huber, Mary 7, 12, 15
Human Research Ethics Board (HREB) 52–3; *see also* Institutional Review Board
human subjects research, *see* research, human subjects
Human Subjects Review Board (HSRB), *see* Institutional Review Board
Hutchings, Pat 7, 12, 15, 23

inertia, *see* pathologies of teaching, inertia
informed consent 52, 54, 55, 88, 102, 161
institutional review board (IRB) 22, 53–5, 88, 140; *see also* Human Research Ethics Board
International Commission on Mathematics Instruction 1
International Society for the Scholarship of Teaching and Learning (ISSOTL) v, vii, 9, 133, 136, 139
interview 86–91
  conducting an 88–9
  open 87
  preparing for 87
  semi-structured 75, 87, 125
  structured 87
  types of questions 87
  used in STEM education studies 90–1
  writing good questions for 87–8
IRB, *see* institutional review board
ISSOTL, *see* International Society for the Scholarship of Teaching and Learning

junior faculty 6, 11, 17–18

Klein, Felix 1
knowledge survey 41, 68, 75–82
  accuracy of 81–2
  example of 77–9
  used in STEM education studies 82

learning assessment techniques 66
learning
  active 34, 48
  invisible 49
Likert scale, *see* survey, Likert scale
literature search 29–31, 33, 120, 122, 142
  citation chaining 30
  descriptors 29, 30
  ERIC 29–30
  keywords 29, 30
  PsychINFO 29
  search terms 29
  snowball strategy 30
  Web of Science 30

MAA, *see* Mathematical Association of America
Mall, Franklin Paine 1
Mathematical Association of America (MAA) 11, 12, 15, 135, 160
mixed method research design 61, 75
muddiest point paper 65, 121

National Academies 9
National Center for Science and Civic Engagement 9
National Research Council 10, 11, 17, 39
National Science Foundation 10, 40, 134
National Science Teachers Association 10
New Experiences in Teaching Project 11
Nobel Prize 1, 47
normalized gain 46, 48, 49

one minute paper 64, 65
open access 139

INDEX | 177

p-value 46
pathologies of teaching 146–7
  amnesia 146–7
  fantasia 146–7
  inertia 146–7
pedagogical amnesia 2, 147;
  see also pathologies of teaching
Pickering, Miles 1
pilot
  interview 88, 91
  study 35, 46, 120
  survey 69, 72
  test(ing) 72, 117
  think-aloud 104, 105
President's Council of Advisors on Science and Technology 10
prior cohort 43, 44, 51, 133
professional development 11, 146
Project NExT, *see* New Experiences in Teaching Project

qualitative research standards 112, 126–8
  coherent 127, 128
  communicable 127, 128
  inter-coder reliability 127, 128
  justifiable 127, 128
  transferability of themes 127, 128
  transparent 127, 128
quantitative research standards 112, 127
  generalizability 127
  reliability 127
  validity 127
quasi-experimental study 41

random assignment 31, 40, 41, 51, 127
rank and tenure, *see* tenure and/or promotion
Research in Undergraduate Mathematics Education (RUME) 12–15
research
  action-based 2
  design 25, 35, 39–46, 51, 55, 60–1, 74
  disciplinary 2
  discipline-based education, *see* discipline-based education research

education 40, 42, 48, 127, 134, 137
  ethics 42, 44–5, 52–5, 137
  human subjects 22, 51–5, 88
  pedagogic(al) 1, 2, 106
  traditional 2
researchable question
  considering disciplinary factors 32–3, 36
  considering feasibility of 34–5, 36
  considering situational factors 31–2, 33, 36
  grounding a question 28–33, 36
  identifying underlying assumptions 33–4, 36
  narrowing a question 27–9, 36
  develop by searching the literature 29–31
  develop by using the SoTL taxonomy 23–7
response rate, *see* survey, response rate
Richlin, Laurie 7
rubric 64, 98, 112, 113–20
  descriptor 113, 114, 116, 119
  dimensions of 64, 113, 114, 115–17, 118, 119, 120
  examples of 113, 114, 115, 116, 117
  holistic-scale 114
  inter-rater agreement 118
  level of performance 113–19
  norming 117, 120
  used in STEM education studies 119–20
RUME, *see* Research in Undergraduate Mathematics Education

sample size 45, 46, 74, 138
Scholarship Assessed 16
scholarship of teaching and learning (SoTL)
  benefits of 145–8
  collaboration(s) with colleagues 18, 31, 35, 49, 51, 54, 107, 111, 119, 124, 132–4
  collaboration(s) with students 106–7
  definition of 7
  distinguishing from good and scholarly teaching 7–8
  evaluation of 15–17

funding for 134
  interdisciplinary nature of 13, 101, 142–3
  movement 6, 7, 9, 11, 12, 22
  multi-disciplinary nature of 13, 30, 119, 134
  origins of 6–8
  taxonomy, *see* SoTL taxonomy
  valuing 17–18
Scholarship of Teaching and Learning Reconsidered 9
Scholarship Reconsidered 6, 16, 145
scholarship
  of application 6, 8, 15
  of discovery 6, 8, 13, 15, 16
  of engagement 6
  of integration 6, 8, 15
  of teaching 6–7
Shulman, Lee 2, 7, 40, 147
SoTL and DBER 11, 14
  mathematics as a case study 11–15
SoTL taxonomy 14, 23–6, 46, 51, 61, 111, 147
  theory-building 23, 101
  vision of the possible 23;
  see also SoTL taxonomy, What could be?
What could be? 14, 23–5, 41, 51, 61, 66, 75, 82, 91, 96, 105, 147
What is? 14, 23–7, 32, 33, 35, 46, 49, 50, 51, 61, 66, 75, 82, 91, 96, 97, 98, 105, 111–12, 147
What works? 14, 23–7, 32, 33, 35, 40, 41, 48, 49, 50, 51, 66, 75, 82, 91, 96, 98, 105, 111, 140, 147
SoTL, *see* scholarship of teaching and learning
statistical significance 43, 45–6, 47, 50, 51, 68, 74
student affordances 32, 33
student voice(s) 13, 105–7
students as co-investigators 54, 55, 105, 107
survey(s) 43, 50, 51, 54, 61, 62, 63, 64, 66, 68–75, 112, 171
  compared to focus group 92
  compared to interview 87
  design 69–72, 74
  knowledge, *see* knowledge survey
  Likert scale 48, 61, 63, 68, 71, 72, 112

survey(s) (*cont.*)
   open-ended responses 64, 71, 112, 120, 122
   reliability 72, 75, 127
   response rate 73–5
   used in STEM education studies 44, 69, 75, 162–3, 171
   validity 69, 72, 75, 127
switching replications design 42–3; *see also* research, design

t-test 45
taxonomy 27, 28
   Bloom's 27, 47, 82, 121, 172
   of mathematical knowledge and expertise 100–1, 172
   for pathologies of teaching 146–7
   for pitfalls of student learning 147

of significant learning experiences 172
of SoTL questions, *see* SoTL taxonomy
teaching
   commons 12, 15
   as community property 2, 7
   good 6, 8
   reflective 8
   scholarly 6, 8, 17, 18
tenure (and/or promotion) 15, 16, 17, 18, 138, 145, 146
think-aloud 14, 86, 90, 91, 97–105, 107, 120, 121
   concurrent report(s) 98, 101
   conducting a 101–4
   protocol 101, 103, 105
   retrospective report 101, 102
   sample warm-up script for 101–3
   underlying assumption (hypothesis) 97, 105

   used in STEM education studies 97–101
   validity of 97
threshold concepts 31, 32, 50
Thurston, Robert H. 1
transcript(ion) 89–90, 91, 95, 96, 104, 105, 120, 125
treatment group, *see* experimental group
triangulate evidence, *see* triangulation
triangulation 14, 60–1, 66, 76, 82, 107

voice recognition software (VRS) 90
VRS, *see* voice recognition software

Wieman, Carl 1, 47

Milton Keynes UK
Ingram Content Group UK Ltd.
UKHW020615220923
429161UK00004B/253